IMAGES
of America

THE MARYLAND STATE FAIR

Celebrating 125 Years

On the Cover: The Midway is always popular at the fair. Delicious food such as snow cones, cotton candy, and candy apples is available, as well as local favorites such as Maryland crab cakes. (Courtesy Maryland State Fair Archives.)

IMAGES
of America

THE MARYLAND STATE FAIR
Celebrating 125 Years

Paige Horine
on behalf of the Maryland State Fair

ARCADIA
PUBLISHING

ISBN 978-1-5316-2592-4

Published by Arcadia Publishing
Charleston SC, Chicago IL, Portsmouth NH, San Francisco CA

Library of Congress Catalog Card Number: 2006922040

For all general information contact Arcadia Publishing at:
Telephone 843-853-2070
Fax 843-853-0044
E-mail sales@arcadiapublishing.com
For customer service and orders:
Toll-Free 1-888-313-2665

Visit us on the Internet at www.arcadiapublishing.com

To my husband, John, who first held my hand
at the Maryland State Fair.

Contents

ACKNOWLEDGMENTS

There are so many people who have had an integral part in the making of this book. My thanks go to Jason Sellman, photographer extraordinaire; the Baltimore County Public Library; the Baltimore County Historical Society; Jeff Korman and the Enoch Pratt Free Library; and my very patient and encouraging family, John, Devon, and Amelia.

This book would not have been possible if not for the dedicated efforts of the many people who worked to preserve the very special history of the Maryland State Fair. The staff at the office of the Maryland State Fair worked endlessly to compile, archive, photograph, and research the information needed to complete this book.

I wish to thank Grove Miller and the board for their wisdom in allowing this project to go forward. Thanks to Max Mosner for being a calming influence and an incredible resource on the history of the fair that is uniquely his. Thanks to Edie Bernier for bringing me on board and taking many of the spectacular images in this book; Andy Cashman for his input, advice, and endless knowledge; and Rebecca Williams for her patience and being the only person the computers would actually listen to and respect.

I would especially like to thank Laura Janicki, who spent countless hours researching, cross checking, photocopying, and correlating images and information on the fair's history. Without her attention to detail and dedication to the project, this book would not be possible.

INTRODUCTION

It's hard to believe that "the Eleven Best Days of Summer" started as only one day in a place called Ridgley's Grove, not far from where it is held today. The tradition of what is now known as the Maryland State Fair started over 100 years ago. The origin of the tradition began as early as the 18th century. Records indicate that a fair was held on the estate of prominent businessman John Eager Howard in 1745.

Fairs as we know them today were actually started by groups of farmers and businessmen who gathered to form agricultural societies. The goals of the agricultural societies were to educate and inform farmers. The fairs were an opportunity for the farmers to show off their abundant harvest and enjoy various forms of entertainment such as horse racing.

The forerunner of the Maryland State Fair, the Fair at Timonium, was no exception. In 1878, a small but determined group of local businessmen incorporated to form the Agricultural Society of Baltimore County. Samuel Brady, Dickinson Gorsuch, Thomas B. Todd, William B. Sands, Daniel Jennifer, and Col. Benjamin P. Taylor filed the necessary papers at the courthouse in Towson, the county seat.

A main attraction for this fair was horse racing. Horse racing at the time was not confined to Thoroughbred racing alone; Standardbred racing, trotting, and steeplechases drew a great crowd and contributed to the fair-like atmosphere. Horse racing would continue to be a huge attraction and contribute to the success and inception of the fair.

It was determined that a fair would be held in Baltimore County. A likely site was in Timonium, which had hosted races since the early 19th century. It was close to the railroad, which was critical to provide transportation for spectators and horses to the site.

As the site in Timonium was more of a site for horse racing than a true racetrack, purses were offered to entice horsemen to bring their finest horses and come to race. The race dates often only lasted several days, but that was enough to establish an interest in also having a fair.

This small county fair was not officially the Maryland State Fair, nor would it be for many years. Another "state fair" was held at the Pimlico Race Track. Run and managed by John Merryman, it was held in conjunction with Pimlico's racing dates.

There was much competition between the two fairs, and all the fairs in the region that were vying for the name "The" Maryland State Fair, but the rivalry was put aside when the "Maryland State Fair at Pimlico" (its official name) became homeless. The Maryland Jockey Club took over the racing dates at Pimlico, therefore eliminating the attraction of racing from that fair.

Without the draw and allure of Thoroughbred racing, the Maryland State Fair was forced to find a new location. The decision was made to hold joint fairs. In 1894 and 1897, the Maryland State Fair and the Agricultural Society of Baltimore County joined together to hold joint fairs in Timonium.

The joint fairs at Timonium continued to be a tremendous success year after year. Horse racing continued to be the main draw, but the fair expanded to include much more. The influence of the agricultural society could be seen, as the fair continued to be a place for farmers to exhibit their abundant harvest and finest livestock.

The fair continued through the decades, growing larger and more popular every year. It managed to change with the times, staging a very popular Airmail Delivery in 1918, slowing down during World War I, and offering the Beautiful Baby contest, which started during the 1920s.

The fair continued in its popularity, led by Matthais L. Daiger. In 1943, for the first time in 65 years, the fair would not be held. The fair, like the rest of America, was focusing its efforts on World War II. Horses would not run, cattle would not be shown, and the pies would have to be enjoyed at home. All around the country, people were asked to make sacrifices, and the people of Baltimore County were no exception. The fairgrounds that once teamed with livestock and spectators would be used as a storage depot and truck repair center for the U.S. Army.

When the war was over, Americans were eager to celebrate, and the fair was a perfect place to show everything that was good in America. America prospered and so did the fair, continuing to grow once again in size and popularity.

Through the 1960s, 1970s, 1980s, and 1990s, right through to the new millennium, the Maryland State Fair has persevered. This book brings to life the history and artifacts that have been painstakingly collected to highlight the fair's remarkable and exciting past. From the thrill of the Midway to the grandeur of Thoroughbred racing and everything in between, the fair is truly "the Eleven Best Days of Summer."

This photograph, taken in 1909, shows a cow-judging contest. Farmhands can be seen in the background waiting to help or just observing the competition. Seen holding the ribbons is Dr. A. C. McCurdy, a resident of Towson and also a dentist. The fair was a time for career and gentlemen farmers alike to show off the fruits of their hard work. Cows were judged then, as they are now, on breed standards, turnout, and overall health. In a time when it was a challenge at best to advertise prime livestock to supplement a farm's income, a prize-winning cow was a good endorsement for a farmer's stock. The farmers often made the long journey to the fair not just for the fun and adventure, but also for the chance to sell livestock at premium prices. The popular breeds at this time were Holstein, Jersey, and Devon. (Courtesy Baltimore County Public Library.)

One

THE EARLY YEARS

An abundant harvest is shown in this 1912 photograph. The state fair was an important outing worthy of getting dressed up. Men wore straw boater hats and bow ties, which were the fashion of the time. The fair was an attraction for people from all walks of life—not just farmers anxious for an outing but also members of Baltimore society often came to cheer on their favorite racehorse. (Courtesy Baltimore County Public Library.)

The Maryland State Fair has had many names throughout its 125-year history. In the early years, there were several fairs throughout the state that bore the name of the Maryland State Fair. The two most prominent were the one held in Timonium by the Agricultural Society of Baltimore County and the State Fair held at Pimlico Race Course. In 1894, the two merged to hold one joint fair. (Courtesy Maryland State Fair Archives, photographed by Jason Sellman.)

Pictured here is the board of the Agricultural Society of Baltimore County, led by Alexander McCormick (center), who was president of the society from 1902 to 1918. (Courtesy Maryland State Fair Archives, photographed by Jason Sellman.)

This is a photograph taken of the founding fathers of the Agricultural Society of Baltimore County. The picture was taken outside of the Towson Courthouse on Chesapeake Avenue. The board was made up of local businessmen and farmers who wanted to see an annual fair in Baltimore County become a reality. (Courtesy Maryland State Fair Archives, photographed by Jason Sellman.)

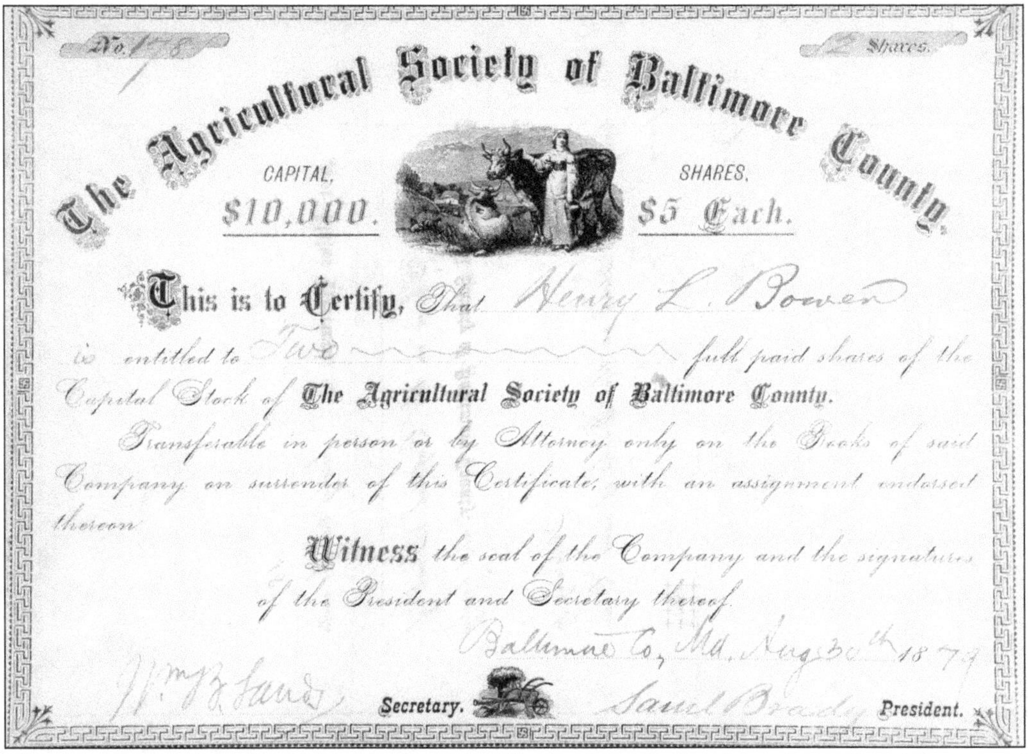

No. 178 — 2 Shares.

The Agricultural Society of Baltimore County

CAPITAL, $10,000. SHARES, $5 Each.

This is to Certify, That Henry L. Bowen is entitled to Two full paid shares of the Capital Stock of The Agricultural Society of Baltimore County.

Transferable in person or by Attorney only on the Books of said Company on surrender of this Certificate, with an assignment endorsed thereon.

Witness the seal of the Company and the signatures of the President and Secretary thereof.

Baltimore Co, Md, Aug 30th 1879

Wm. B. Sands, Secretary. Saml Brady, President.

This certificate of stock issued in 1879 shows the corporate title of the Agricultural Society of Baltimore County. This shows the first step in the evolution of the official name of the Maryland State Fair. In 1906, the organization was again incorporated into the Maryland State Fair and Agricultural Society of Baltimore County, the beginning of the Maryland State Fair of today. (Courtesy Maryland State Fair Archives, photographed by Jason Sellman.)

Take lists for judges books from this

PREMIUM LIST

32D ANNUAL

MARYLAND STATE FAIR

MARYLAND STATE FAIR AND AGRICULTURAL SOCIETY OF BALTIMORE CO. MD. SEAL INCORPORATED 1878.

..AND..

Agricultural Society of Baltimore Co.

TIMONIUM, N. C. R. W.

SEPTEMBER 6, 7, 8, 9, 10.

1910

This premium list shows the Maryland State Fair and Agricultural Society of Baltimore County, Timonium. This name evolved into what is known today as the Maryland State Fair. The fair, still run and organized by the Agricultural Society of Baltimore County in 1910, would not be known officially as the Maryland State Fair until the late 1930s. Total premium prize money given out in 1910 was close to $20,000, quite a good incentive to attract exhibitors. (Courtesy Maryland State Fair Archives, photographed by Jason Sellman.)

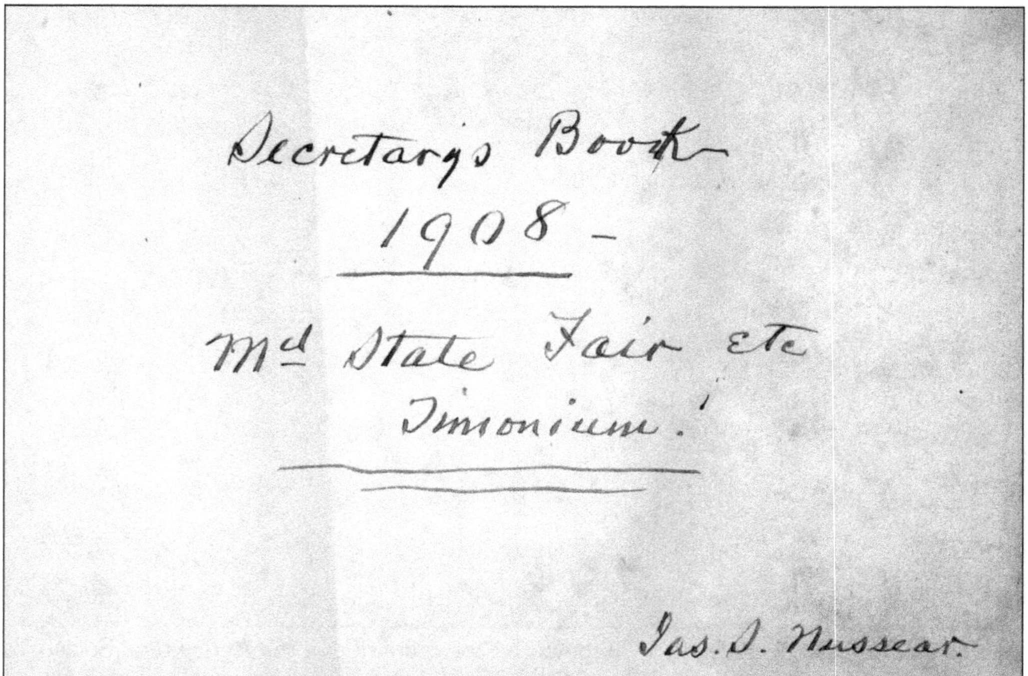

Secretary's Book

1908 –

Md State Fair etc

Timonium!

Jas. S. Nussear.

This book from Jason Nussear, the fair secretary, shows his handwritten notes for the 1908 fair. The notes include entries to the fair as well as premiums paid and other expenses incurred. (Courtesy Maryland State Fair Archives, photographed by Jason Sellman.)

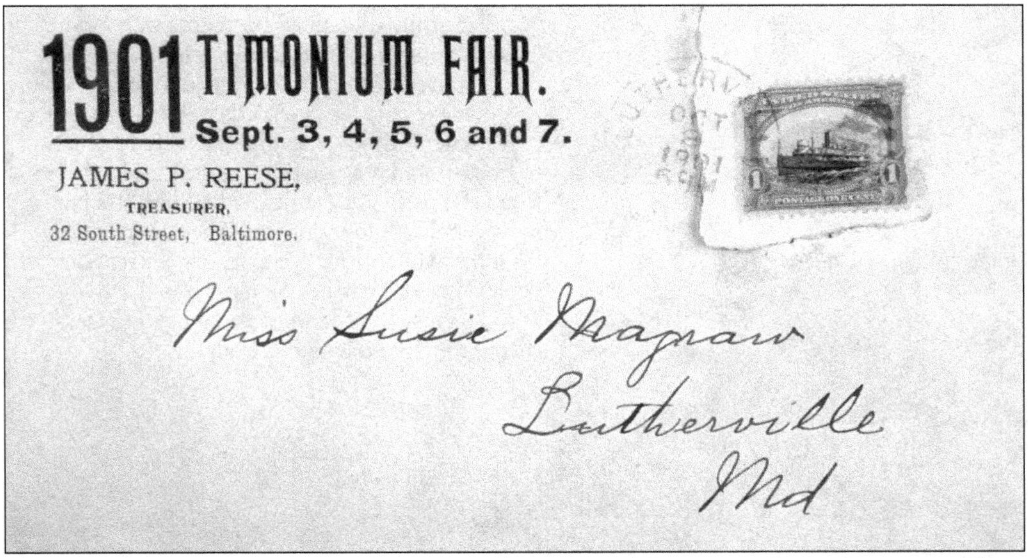

The merger of the two fairs can be seen in this postcard from 1901. It was not a smooth or straightforward transition but a merger driven from necessity. Speculation of financial difficulties, mismanagement, and lost racing dates all drove the merger. (Courtesy Maryland State Fair Archives, photographed by Jason Sellman.)

In 1914, the fair was up and running under the name of the Maryland State Fair and the Agricultural Society of Baltimore County, held in Timonium. A premium list is a book that includes all the classes that could be entered at the fair. All livestock, home arts, produce, and specialty classes were listed, as well as the premiums, or prize money, to be paid. (Courtesy Maryland State Fair Archives, photographed by Jason Sellman.)

Beautifully illustrated postcards were often sent out to save the date or announce when the racing dates would be held for a particular course. Souvenir postcards were also available for sale to finance airmail deliveries, which were popular attractions at fairs and exhibits in the early 20th century. In 1880, the fair was held for four days and there were 350 stalls available for competitors, with about 100 held for cattle, the most popular livestock. (Courtesy Maryland State Fair Archives, photographed by Jason Sellman.)

This postcard advertising the Maryland State Fair depicts a beautiful rooster. Poultry, including ducks, geese, turkeys, and pigeons, were always popular attractions at the fair. It was a great way to see new and ornamental breeds as well as excellent examples of birds that could be used on the farm. (Courtesy Maryland State Fair Archives, photographed by Jason Sellman.)

This adorable cowgirl advertises the Maryland State Fair. Postcards like this were available at the fair as souvenirs. People often bought them and mailed them for a penny to friends from the fair, often bragging about the wonderful time that was enjoyed at the fair. (Courtesy Maryland State Fair Archives, photographed by Jason Sellman.)

According to a September 4, 1882, article, athletic games were also held on the grounds during the fair. It was a time for local athletic and rowing clubs to come out and show their strength. This type of postcard was used to encourage and entice participants. (Courtesy Maryland State Fair Archives, photographed by Jason Sellman.)

This 1908 photograph shows another fair participant who arrived by horse-drawn carriage. The carriage is a four-wheel cart pulled by a Standardbred who, in the driving fashion of the day, has a bobbed tail. His harness, featuring an over-check, is also in fashion. An over-check is the thin piece of leather that goes from the bridle to the saddle and prevents the horse from putting his head down far enough pull the reins out of the driver's hands. (Courtesy Baltimore County Public Library.)

Another view of the carriage, interestingly enough, shows that the leather harness does not include breeching straps, thin straps of leather that would go around the horse's rump and prevent the carriage from riding up onto the back of the horse. This could mean that either the cart has brakes, a more modern invention, or that the carriage was only used for relatively short distances on flat ground. (Courtesy Baltimore County Public Library.)

Two lovely ladies, Marjorie Brooks McKee (center) and Helen Tarbert Linville (holding a parasol), sit in a four-wheel carriage. The ladies were probably escorted to the fair by a groom or male family member who drove the carriage, as neither lady is dressed in the traditional gloves and lap robe one would wear to protect hands and clothes while driving. Dr. Alan Brooks sits astride his horse and looks upon the scene. (Courtesy Baltimore County Public Library.)

Mrs. Cull, wife of attorney Rodger W. Cull, is shown with her horse Punch, harnessed to a four-wheel dogcart. In the early 20th century, it was very common for people to arrive at the fair via horse-drawn carriage. The fair was also conveniently located next to North Central Railroad, which would bring up to nine carloads of people a day to the fair. (Courtesy Baltimore County Public Library.)

The front of the judges' box and grandstands are shown in this 1904 photograph of the Fair at Timonium. The dirt track would accommodate Thoroughbred, Standardbred, and steeplechase racing. (Courtesy Maryland State Fair Archives, photographed by Jason Sellman.)

This 1906 racing program shows yet another name the fair was known by: "the Great Timonium Fair." The program may likely have been highlighting Thoroughbred racing, but trotters and steeplechases also offered opportunities for betting. (Courtesy Maryland State Fair Archives, photographed by Jason Sellman.)

Mr. Biggs stands at the ready, waiting for his race to begin. Standardbred racing was every bit as popular with the crowd as Thoroughbred racing. It was popular with the owners too, as the jockeys or drivers for the Standardbred horses did not have to be as light and small as the Thoroughbred jockeys, allowing many of the Standardbred owners to drive and race as well. (Courtesy Baltimore County Public Library.)

Thoroughbred racing wasn't the only form of racing found at the fair. Standardbred racing was a thrilling and popular sight that drew thousands of spectators. This large field of eight horses emphasizes that there was plenty of interest among the trotter enthusiasts as well. (Courtesy Baltimore County Public Library.)

This postcard from the early 1900s shows the large wooden grandstands, a neat white picket fence, and a perfectly groomed track at Timonium. The tall mature trees that lined the racetrack offered relief from the summer sun for weary fairgoers or race fans. (Courtesy Maryland State Fair Archives, photographed by Jason Sellman.)

The stands sit quietly waiting for the next race. In 1892, according to the *Baltimore County Union*, "improvements will be made on county fairgrounds totaling $4,000. These will include improved box and cattle stalls, roofing, fencing and other betterments that will add greatly to the value of the society's property." The grandstands here were part of that renovation. (Courtesy Maryland State Fair Archives, photographed by Jason Sellman.)

Cards such as this 1911 image were sent out from the Maryland State Fair to area patrons and racetracks. Trainers could then include the fair in their racing schedules. There were several area fairs, such as York and Frederick Counties, that made up a racing circuit at the time. Timonium's close proximity to the North Central Railroad made it an easy destination to reach. (Courtesy Maryland State Fair Archives, photographed by Jason Sellman.)

For 10¢, one could get an afternoon of fun in 1913. The race program of the Maryland State Fair and Agricultural Society of Baltimore County, held in Timonium, promises great fun and possible riches. (Courtesy Maryland State Fair Archives, photographed by Jason Sellman.)

Pictured are souvenirs from various years of the Maryland State Fair. The creamer is from 1912, the pitcher from 1911, the tumblers from 1906, and the salt and pepper shaker set from 1911. The souvenirs were given as gifts to friends of the fair, such as board members, and were available for purchase by fair enthusiasts. (Courtesy Maryland State Fair Archives, photographed by Jason Sellman.)

Another popular souvenir, a wooden nickel, is from the 1927 Maryland State Fair. (Courtesy Maryland State Fair Archives, photographed by Jason Sellman.)

This beautiful ribbon, a souvenir of another kind, is an example of the many prizes that awaited an exhibitor at the fair. It is a silk ribbon held by a silver pin in a traditional repoussé pattern. (Courtesy Maryland State Fair Archives, photographed by Jason Sellman.)

This pennant was one of many souvenirs sold at the Maryland State Fair. (Courtesy Maryland State Fair Archives, photographed by Jason Sellman.)

The judges' stand, shown here in 1912, sports advertisements for other local fairs. When the racing dates drew to a close for the Fair at Timonium, race fans had other venues to attend. (Courtesy Baltimore County Public Library.)

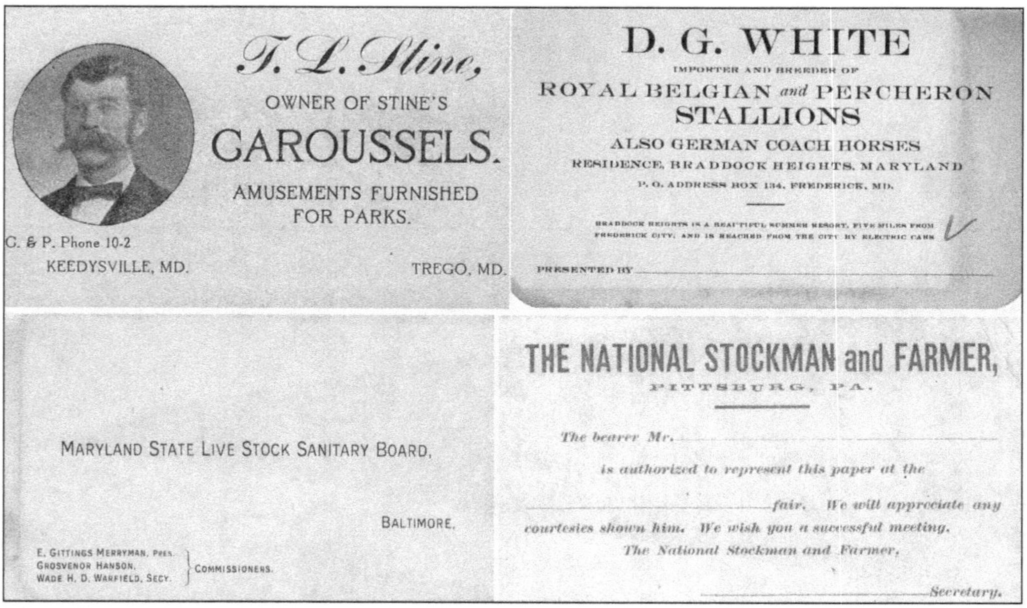

The office of the fair often received—and still does—many pieces of solicitation. Entrepreneurs saw a need and tried to fill it. Mr. T. L. Stine offered carousels, a very popular attraction. (Courtesy Maryland State Fair Archives.)

Any time a crowd gathers, it is a good opportunity to sell something. Merchants began to see the fair as a great opportunity to advertise whatever it was they had to sell. Charles Rudd advertised in the racing program, trying to promote his cars. As he was in Lutherville, fairgoers did not have far to go if they were interested in trading in the horse and buggy for a car. (Courtesy Maryland State Fair Archives, photographed by Jason Sellman.)

Merchants who displayed at the fair also took advantage of the large crowds. Buttons like this from the Old Town Merchants and Manufacturers Association were still a novelty in 1910. Founded in 1883, this organization was formed to promote the interest of the Old Town section of East Baltimore. The fair gave them an opportunity to promote the area and bring issues such as trade and civic involvement to light. (Courtesy Maryland State Fair Archives, photographed by Jason Sellman.)

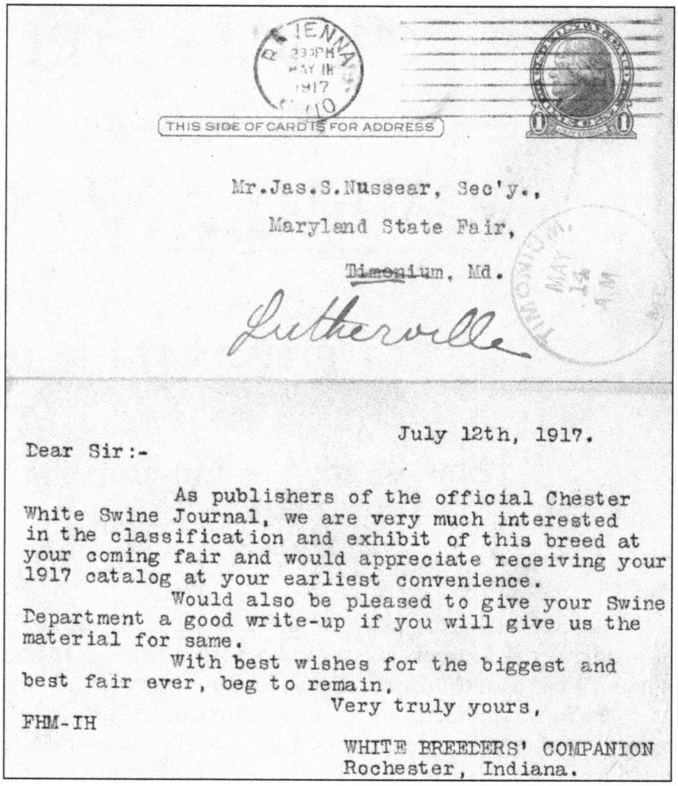

THIS SIDE OF CARD IS FOR ADDRESS

Mr. Jas. S. Nussear, Sec'y.,

Maryland State Fair,

~~Timonium~~, Md.

Lutherville

July 12th, 1917.

Dear Sir:-

As publishers of the official Chester White Swine Journal, we are very much interested in the classification and exhibit of this breed at your coming fair and would appreciate receiving your 1917 catalog at your earliest convenience.

Would also be pleased to give your Swine Department a good write-up if you will give us the material for same.

With best wishes for the biggest and best fair ever, beg to remain,

Very truly yours,

FHM-IH

WHITE BREEDERS' COMPANION
Rochester, Indiana.

The *Chester Swine Journal* writes a compelling letter to secretary Jason Nussear expressing interest in participating in the 1917 fair. The fair was continuing to grow and expand, attracting participants from as far away as Rochester, Indiana. (Courtesy Maryland State Fair Archives, photographed by Jason Sellman.)

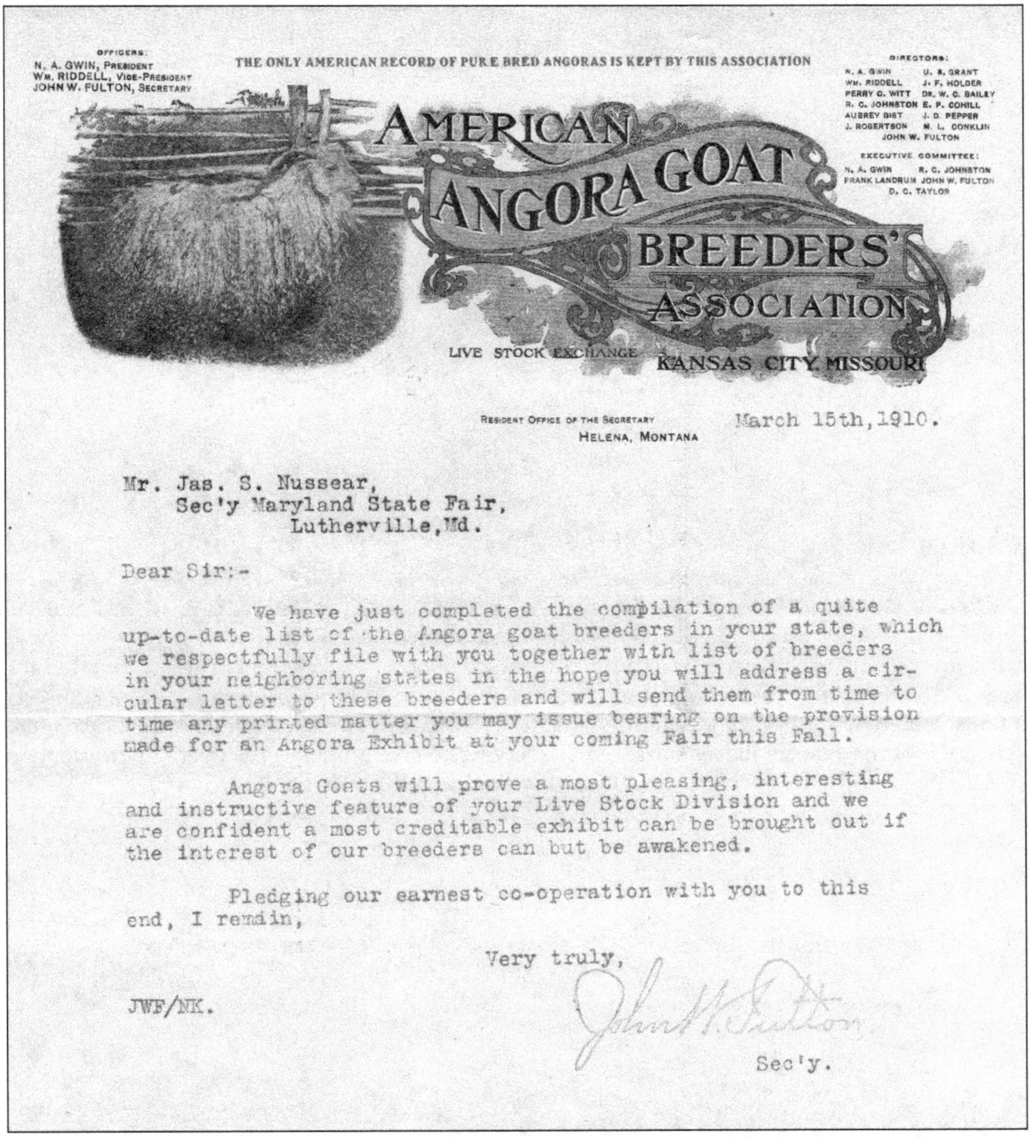

OFFICERS:
N. A. GWIN, PRESIDENT
WM. RIDDELL, VICE-PRESIDENT
JOHN W. FULTON, SECRETARY

THE ONLY AMERICAN RECORD OF PURE BRED ANGORAS IS KEPT BY THIS ASSOCIATION

DIRECTORS:
N. A. GWIN U. S. GRANT
WM. RIDDELL J. F. HOLDER
PERRY O. WITT DR. W. C. BAILEY
R. C. JOHNSTON E. P. COHILL
AUBREY BIST J. D. PEPPER
J. ROBERTSON M. L. CONKLIN
JOHN W. FULTON

EXECUTIVE COMMITTEE:
N. A. GWIN R. C. JOHNSTON
FRANK LANDRUM JOHN W. FULTON
D. C. TAYLOR

AMERICAN ANGORA GOAT BREEDERS' ASSOCIATION

LIVE STOCK EXCHANGE KANSAS CITY, MISSOURI

RESIDENT OFFICE OF THE SECRETARY
HELENA, MONTANA

March 15th, 1910.

Mr. Jas. S. Nussear,
Sec'y Maryland State Fair,
Lutherville, Md.

Dear Sir:-

We have just completed the compilation of a quite up-to-date list of the Angora goat breeders in your state, which we respectfully file with you together with list of breeders in your neighboring states in the hope you will address a circular letter to these breeders and will send them from time to time any printed matter you may issue bearing on the provision made for an Angora Exhibit at your coming Fair this Fall.

Angora Goats will prove a most pleasing, interesting and instructive feature of your Live Stock Division and we are confident a most creditable exhibit can be brought out if the interest of our breeders can but be awakened.

Pledging our earnest co-operation with you to this end, I remain,

Very truly,

John W. Fulton

Sec'y.

JWF/NK.

In 1910, the fair's growing popularity was gaining national attention. In this letter from John W. Fulton, secretary of the America Angora Goat Breeders Association, to Jason Nussear, secretary of the Fair in Timonium, Mr. Fulton is enlisting the help of the fair. The increasing number of people attending the fair created an ideal opportunity to promote a new or unusual breed of livestock such as the Angora goat. By sending out literature from the Goat Breeders Association to local goat breeders, the fair accomplished two things: attracting new exhibitors from Kansas City, Missouri, and increasing attendance among local goat breeders. (Courtesy Maryland State Fair Archives.)

Double reins, heavy bits, and bobbed tails show that these horses were used to going fast. These riders are waiting their turn or watching how the competition takes the last fence. In a hunter pace or hunter trials, riders must clear natural fences such as brush and timber. The course is often small enough to be viewed from one vantage point but large enough that the horse must gallop the entire course. Judges stationed at every fence give points to the overall performance of each horse and rider at that fence. (Courtesy Baltimore County Public Library.)

Riders line up, waiting anxiously for a class to be pinned as white-coated stewards look on from the judges' tent. Events such as this jumping contest were popular and common at the Timonium Fair. Local hunt clubs such as the Greenspring Hounds and the Elkridge Hounds welcomed the opportunity to compete at such lovely fairgrounds. (Courtesy Baltimore County Public Library.)

Riders mill about, waiting their turn to compete in what is probably a hunter trial or hunt pace, given the large brush jump in the background. Riders can be seen in the traditional hunt attire, including stock ties. White armbands around the riders' right forearms may have had numbers to identify them to the judge or they denoted a hunt club affiliation. (Courtesy Baltimore County Public Library.)

Riders and their horses line up after a class, waiting for ribbons to be awarded or "pinned." When everyone is finished competing in a riding class, whether in the ring or over fences, the judge will call for the competitors to "line-up in front of the ring steward." The riders' numbers are then called out in order of placement, and trophies or ribbons are given to the competitors. (Courtesy Baltimore County Public Library.)

Bloomfield Farm sets up camp during the Maryland State Fair. Mr. Frank Hoen of Glyndon stands in a shady spot under a tree with his helpers and the cattle in the background. Fair participants, then as now, can be at the fair for a few days or all week. They try to make this workspace as comfortable as possible, as showing cattle is hot, hard work. (Courtesy Baltimore County Public Library.)

Grapes hang on an arbor in this display at the fair. Excellent examples of different varieties of grapes could be viewed, which might inspire someone to try a new variety in their own garden or vineyard. (Courtesy Baltimore County Public Library.)

Exhibitors dressed in their finest hats, suits, and dresses proudly stand by their cows, who are also turned out without a speck of dust or dirt. The finest of their day, these cows show a marked difference from their modern-day counterparts. The cows that are produced today are significantly larger than their ancestors from 1909. (Courtesy Maryland State Archives, photographed by Jason Sellman.)

This 1909 photograph shows Milton Easton of Ellicott City and his pony Bobby Alone. Bobby appears to have won a blue ribbon in an in-hand class. An in-hand class is one in which the horse or pony is not ridden and is shown by a handler leading him by the halter, bridle, or long-lines, as is shown with Bobby Alone. This class allows the judge to see the animal's natural way of going without the influence of a rider. (Courtesy Baltimore County Public Library.)

This massive steam engine from the North Central Railroad played an important role in the success of the fair. The NCRW twice caused a relocation of the fair when it expanded its line, running right through the racecourse. When the fair finally moved to its present location on the corner of Timonium and York Roads, the NCRW became the fair's savior. Bringing as many as nine passenger trains a day, the railroad provided convenient transportation for those who wanted to attend the fair. The train made traveling from Baltimore City or Pennsylvania a reality. (Courtesy Baltimore County Public Library.)

This train schedule for the NCRW included dates, times, and prices for transportation to the fair. Livestock and racehorses also found easy transportation to the fair via the train. This helped boost not only spectators but exhibitors as well. (Courtesy Maryland State Fair Archives, photographed by Jason Sellman.)

BelleField Mansion, later renamed Timonium Mansion by Sarah Buchanan, is shown in this photograph. Legend has it that after the death of her husband, Mrs. Buchanan visited a friend, Cassandra, who lived in New Orleans. Cassandra decided to return to Baltimore on a later ship. She was never heard from again. Many years later, a pirate named Dominick Lafite lay on his deathbed and confessed to "making a Maryland lady walk the plank." It is said that in her depression after suffering these terrible losses, Mrs. Buchanan decided to rename the beautiful estate Timonium either after Timon of Athens, the melancholy philosopher immortalized by Shakespeare, or after the tower near Alexandria, Egypt, to which Mark Anthony returned when all his friends deserted him. Timonium Mansion was a very prominent estate for centuries and stood on the corner of West Timonium and York Roads. When it was later converted into an inn, it became a haven for travelers due to its close proximity to the North Central Railroad. (Courtesy Baltimore County Public Library.)

Traveling magic acts and games of skill and chance were very popular at the fair. This 1909 scene shows that the Fakir's Booth was no exception. These displays provided a diversion from everyday chores or a just a chance to see something new. Several observers can be seen looking on with interest or cautious speculation. (Courtesy Baltimore County Public Library.)

Crowds stroll through the Midway as a nearby merchant sets up his banners. Merchants often traveled from fair to fair, an excellent way to advertise and sell their wares. Merchandise ranged from household goods to beauty products to medicine. (Courtesy Baltimore County Public Library.)

Ladies and gentlemen in hats and suits stroll the Midway at the Timonium fair in this photograph from 1908. Then as now, the structures are temporary and easily dismantled to follow the fair circuit in the next county or state. (Courtesy Baltimore County Library.)

In 1919, the Midway at the fair had expanded to include many different types of attractions, such as the sideshow attraction "Bowery After Dark." The carousel, a favorite with children and adults, can be seen on the right. (Courtesy Baltimore County Public Library.)

A common ride on the Midway, the Whip was just as popular in 1908 as it is today. It took many hours of hard work to set up the rides, but countless adults and children thoroughly enjoyed the results. In the days before electricity, the rides were powered by steam or by strong young men who turned the carousel with a hand crank. (Courtesy Baltimore County Public Library.)

Though not as large as today, the Midway in 1909 was still a popular attraction for those who attended the fair. The Midway is completely mobile, as it was then. There are no permanent structures devoted to the Midway at the fair. Games and rides are dismantled and packed away at the end of the fair to travel to another destination. (Courtesy Baltimore County Public Library.)

Another example of entrepreneurs at the fair, roving photographers using tubular cameras would stroll the fairgrounds and entice people into having their pictures taken as souvenirs from the fair. In this picture, a mother nervously looks on as her two lovely daughters have their picture taken for 5¢. (Courtesy Baltimore County Public Library.)

This postcard from 1910 was probably sent to area athletic clubs. Clubs would often hold competitions on the fairgrounds during the fair, as the grounds were suitable and fair time would ensure a large crowd of spectators. (Courtesy Maryland State Fair Archives, photographed by Jason Sellman.)

The first 60 years of the Maryland State Fair were a time of ambition, growth, and exploration. There were many years of uncertainty as to the fate of the fair, but a dedicated group continued to persevere to ensure its success and continuation. The first presidents of the fair were Samuel Brady (1878–1883), Col. Walter S. Franklin (1883–1885), Col. Dennis M. Matthews (1885–1891), Duane H. Rice (1891–1892), Gittings Merryman (1892–1895), Frederick von Kapff (1895–1898), Samuel M. Shoemaker (1898–1900), George W. Yellot (1900–1902), Alexander McCormick (1902–1918), Louis McLane Merryman (1918–1922), Clinton Levering Riggs (1922–1923), and John M. Dennis (1923–1936). These men displayed dedicated and unwavering leadership in taking the fair to greater heights of popularity and making the fair what it is today. (Courtesy Baltimore County Public Library.)

Two

I Remember When . . .

As the fair continued, the role of women in the kitchens of America was also changing. It was feared that the entries in the home arts department at the fair would diminish. Women wanted to get out of the kitchen and into the modern world. An article in a 1920s newspaper commented that several "bobbed hair flappers" had been seen baking cakes at the fair. The fair continued to be popular with women and men from all walks of life into the 1920s, 1930s, and 1940s. This young woman presents a platter of exquisite grapes complete with a ribbon won for exceptional quality. (Courtesy Maryland State Fair Archives.)

A display of honey shows many different varieties produced in Maryland. According to their Web site, the Maryland Beekeepers Association is "dedicated to the advancements of beekeeping and the improvement of beekeepers skills in Maryland since 1908." The Maryland Beekeepers Association exhibition can still be seen at the fair today. (Courtesy Maryland State Fair Archives.)

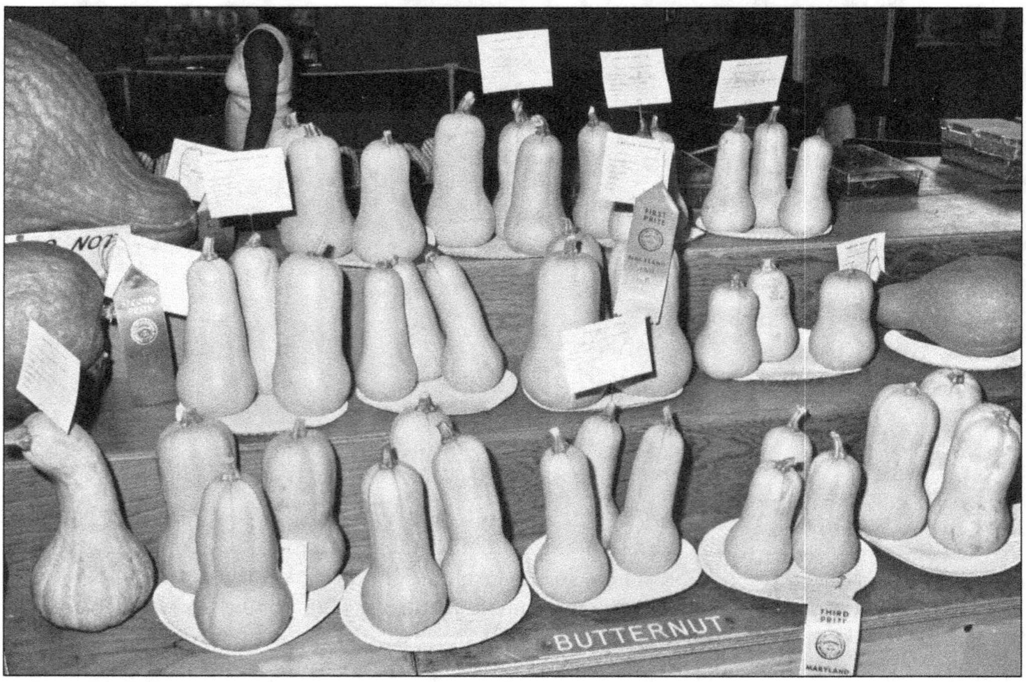

An abundance of prize-winning squash is seen here. Squash need plenty of water and warm sunny temperatures to grow to these proportions. They can be grown in any house or farm garden or even in a barrel on a terrace or balcony of an apartment. (Courtesy Maryland State Fair Archives.)

Anyone who lives in Maryland knows that summer is synonymous with tomatoes. Almost everyone with a square inch of earth to spare has at least one tomato plant, which with the right conditions will produce an endless supply of juicy red tomatoes that can be enjoyed on sandwiches, salads, or just by themselves. (Courtesy Maryland State Fair Archives.)

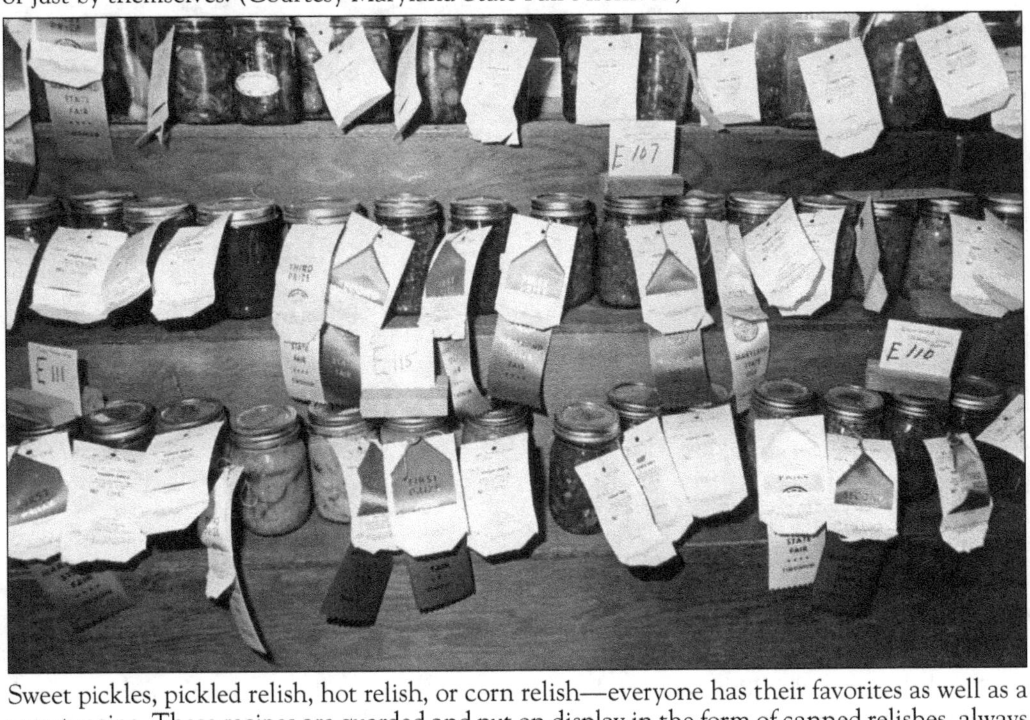

Sweet pickles, pickled relish, hot relish, or corn relish—everyone has their favorites as well as a secret recipe. These recipes are guarded and put on display in the form of canned relishes, always a division with a lot of entries at the fair. (Courtesy Maryland State Fair Archives.)

The 1941 Maryland Dairy Judging team is seen here. From left to right are coach H. C. Barker, Charles Stiles from Montgomery County, Timothy Magness and Thomas Clark, both from Harford County, and William King of Montgomery County. (Courtesy Maryland State Fair Archives.)

The 4-H Club has always had a strong presence at the fair. The club started over 100 years ago, around the start of the 20th century. The idea came from the desire to make public-school education more connected to country life and to involve youth living in rural area in civic and community activities. (Courtesy Maryland State Fair Archives.)

A young boy receives a trophy for his efforts. A lot of hard work and care went into preparing this steer for the fair. The fair is an opportunity to show off the yearlong effort. (Courtesy Maryland State Fair Archives.)

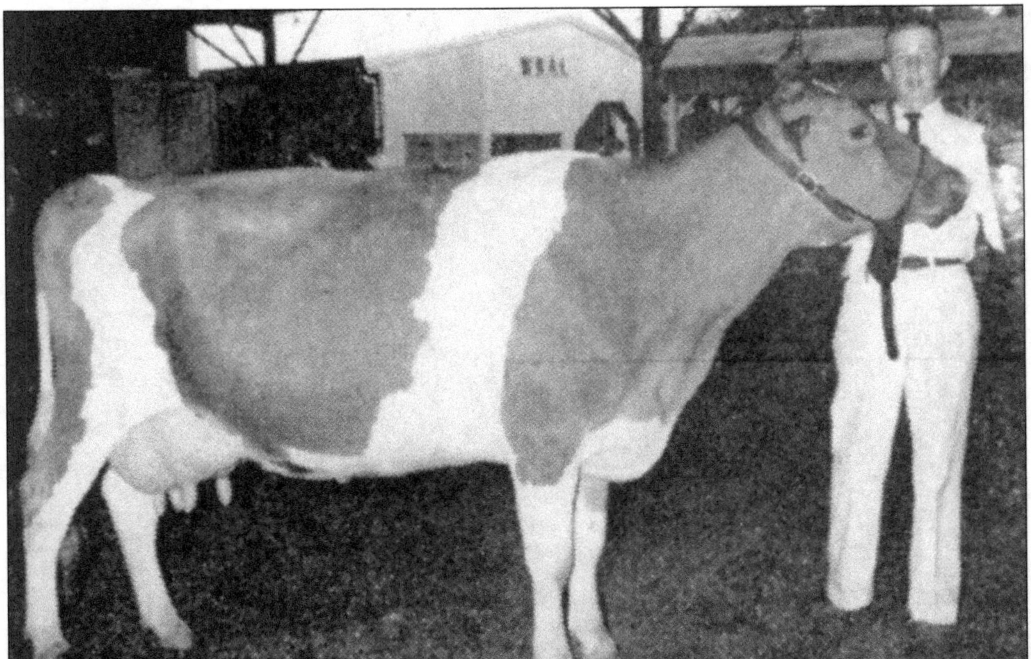

George Debman of Upperco, Maryland, is shown here with his grand champion Guernsey cow. This year, 1954, is the second year he won this award. George certainly lived up to the 4-H motto "to make the best better." (Courtesy Maryland State Fair Archives.)

Spencer Street of Forest Hill is shown here with his 4-H beef champion Hereford steer. The 4-H Club is a community of young people ages 8 through 18 across America. It is the largest out-of-school activity in the United States. (Courtesy Maryland State Fair Archives.)

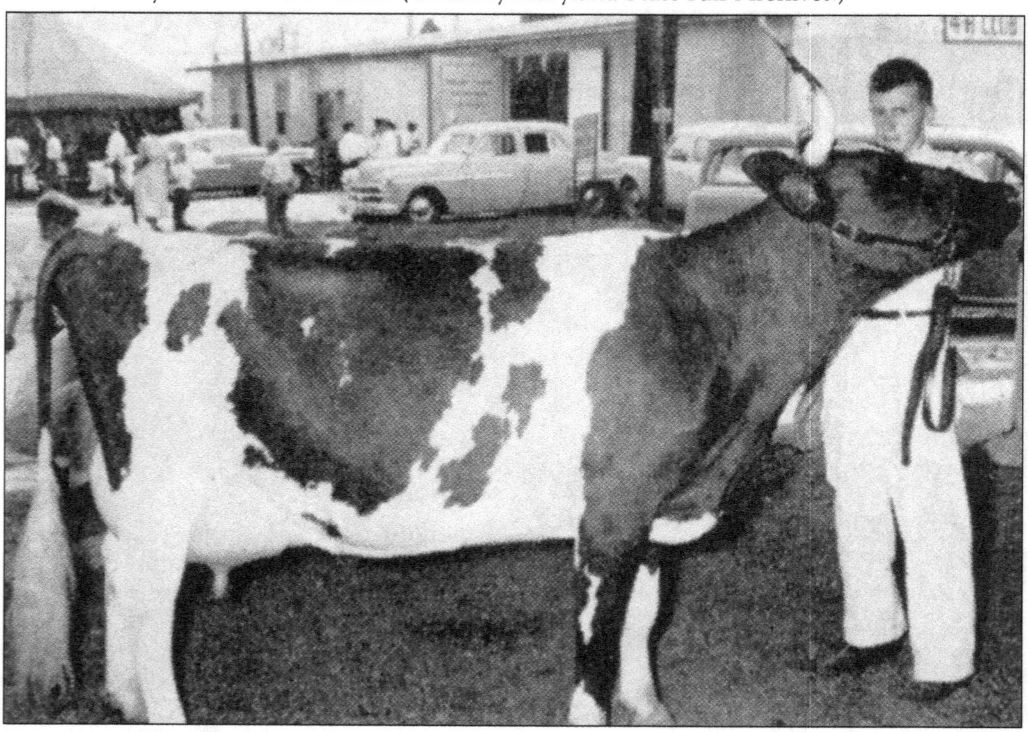

James Ray Hill of Woodbine shows off his grand champion Ayrshire cow. (Courtesy Maryland State Fair Archives.)

Judging cows is an art that takes years to learn and perfect. There are many different attributes of each breed that must be taken into consideration when deciding on a winner. To make the process less complicated, a judging card is often used; this is a card preprinted with desired traits of the breed. The judge can then make his selection based on which animals in the ring best fit the description. (Courtesy Maryland State Fair Archives.)

Criteria for judging dairy cattle are very different from judging beef cattle. When judging dairy cattle, according to the *Livestock Judging Handbook* by Nordby-Beeson and Fourt, "a cow in milk should give general impression of milkiness and display evidence of feminine refinement." This lovely cow looks like a winner. (Courtesy Maryland State Fair Archives.)

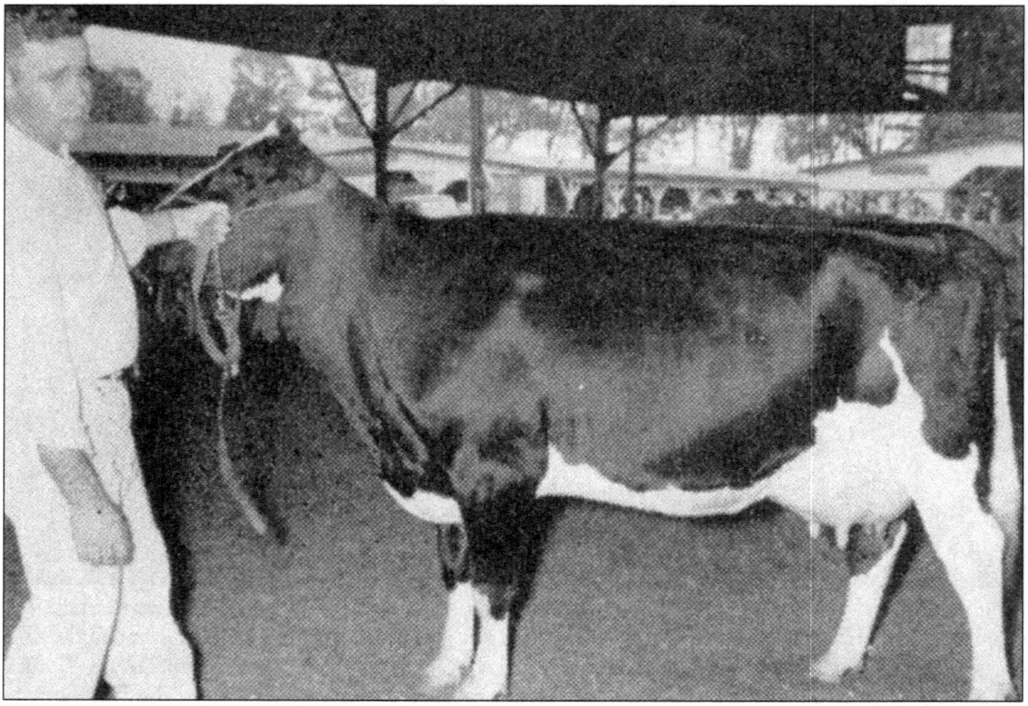

David Brauning of Finksburg, Maryland, is shown here with his Holstein. This animal was a winner in the three-year-old-and-under class. (Courtesy Maryland State Fair Archives.)

Edward Steward (left) of Harford County, shown with an Angus steer, and Harry Lee Eyre of Howard County, shown with a Hereford steer, display their winning entries. (Courtesy Maryland State Fair Archives.)

Edwin C. Fry of Montgomery County is seen receiving a fitting and showing achievement trophy from Dr. K. L. Turk. Fitting and showing is a division at the fair that displays the actual showing of the animal. The emphasis is on the handling, turnout, and showmanship of the handler as opposed to the performance of the animal. (Courtesy Maryland State Fair Archives.)

The crowd looks on with appreciation at a fine beef cow. A prize-winning entry must posses "a short deep body, low set, blocky formed and wide chested, square rump and full quartered individual with a short neck and broad head, characteristic of the breed." (Courtesy Maryland State Fair Archives.)

This 4-H champion female Berkshire swine is exhibited by James Martin of Hampstead. He displays the blue ribbon on the pig's back. There are basically two types of pigs: the purebred and market breed. The Berkshire, Chester White, Duroc, Hampshire, Tamworth, Yorkshire, Poland China, and spotted Poland China are all examples of breeds that are shown at the Maryland State Fair. (Courtesy Maryland State Fair Archives.)

William Laudenklos Jr. of Baltimore stands with his 4-H champion female Yorkshire pig. (Courtesy Maryland State Fair Archives.)

A group of 4-Hers is seen here demonstrating their original project. Founded over 100 years ago, 4-H was a way for youth to learn by doing. The USDA saw that adults in the farming communities were slow to accept new ideas and advancements in farming. The youth, however, were quick to pick up and embrace new ideas and concepts. It was thought that if the youth could become involved in learning, they would pass this information on to the adults. The 4-H organization is an international movement today. (Courtesy Maryland State Fair Archives.)

Ribbons for exhibitors of poultry were bountiful. The fair was an opportunity for spectators not only to see fine examples of common poultry often found on farms but also to see new breeds that would perhaps have larger egg-laying capacity, were heavier, and sometimes were just ornamental. (Courtesy Maryland State Fair Archives, photographed by Jason Sellman.)

Festive banners drape the Exhibition Hall at the fair. Poultry was a very popular exhibit that drew a lot of spectators. In 1937, H. S. Thompson was superintendent of the poultry division at the fair. Nearly 2,000 ducks, geese, turkeys, and pigeons were in cages ready to be shown. (Courtesy Maryland State Fair Archives.)

William Hood (left) and Amos Keller (right), both of Frederick, represent 4-H with a poultry exhibit. The four-leaf clover is the symbol of the 4-H Club. The green of the clover, nature's most common color, represents youth, life, and growth, while the white symbolizes purity. The four *H*'s are Head, Heart, Hands, and Health. (Courtesy Maryland State Fair Archives.)

Sheep have always played an important role on the farm and at the fair. The art of judging sheep takes considerable study and dedication. There are many different breeds, each with different characteristics such as body and wool type. A knowledgeable judge must devote a lot of time to learning this trade. Time must be spent observing different breeds of sheep, their wool types, and the specific purpose for which the sheep is bred, such as wool, market, or breeding stock. (Courtesy Maryland State Fair Archives, photographed by Jason Sellman.)

This photograph shows one entry in a flock class. The sheep in the flock do not have to be related but must have a male and three females of varying ages. The breeder usually chooses the best of his flock to put together a winning pen. The flock featured here are Southdowns. (Courtesy Maryland State Fair Archives, photographed by Jason Sellman.)

Joseph Watcher (left) and James Page of Frederick County are shown here in 1941 with their first-place trophy for Draft Horse Fitting and Showing. In 1941, its 65th year, the fair was going strong and was as popular as ever. Over 50,000 people attended the fair on the opening day. Laban Sparks was president, and Mabel Nusbaum of Carroll County was crowned queen of the Timonium Fair. (Courtesy Maryland State Fair Archives, photographed by Jason Sellman.)

This 1947 photograph shows a blue-ribbon-winning mare and foal. The group is shown in the grass show ring in front of the old horse barns. This picture highlights the fact that the fair is often a family affair in more ways than one. It takes a lot of work to show a horse and her young offspring, and many hands are needed; often the whole family gets involved. (Courtesy Bob Shirley.)

In the late 1930s and early 1940s, the fair was doing so well that improvements needed to be made to keep up with the demands of the crowds. In 1934, $30,000 worth of premiums was given out, excluding racing purses. In 1941, $85,000 worth of improvements was made on the fairgrounds. These improvements included a new inside rail for the racetrack, cattle barns, and an exhibition hall, as well as a new pony show ring. (Courtesy Maryland State Fair Archives.)

Cows were a tremendous draw for attracting exhibitors to the fair. In 1941, Gainville H. Hibberd was in charge of the dairy department, with William R. Harvey in charge of beef cattle. This new building was erected to accommodate the hundreds of cows that attended the fair each year. (Courtesy Maryland State Fair Archives.)

The secretary's office is seen here. A new 250-foot-by-90-foot exhibition hall was constructed to accommodate exhibitions such as University of Maryland Farm Bureau, Maryland Training School for Boys, Maryland State Games and Inland Fish Commission, Women's Handiwork, puppet shows, a cake baking contest, and a homemaking contest. (Courtesy Maryland State Fair Archives.)

Thoroughbred racing was still an important attraction at the fair. Barns were improved to house the hundreds of horses that came in for racing day. Thoroughbred racing, trotting, and steeplechases were all popular among race fans. About $40,000 in purses was paid out to horse race winners in 1941. (Courtesy Maryland State Fair Archives.)

The Office of the Superintendent shows a roomful of women with only one boy. This is a far cry from the early days, when the fair was run only by men. A big sensation was caused in the early 1920s, when K. E. Braithwaite was the first woman elected to the Board of Visitors and Governors for the Baltimore County Agricultural Society. Miss Braithwaite was the manager of the women's department of the fair and also a supervisor of home economics in Baltimore County Schools. Miss Braithwaite's appointment even warranted attention from the local newspaper, which commented that a member of the "fair sex" was elected to the board. (Courtesy Maryland State Fair Archives, photographed by Jason Sellman.)

Maryland once had hundreds of thriving and abundant tobacco farms. Tobacco was a very important crop to the state. With the decrease in the popularity of smoking today, most tobacco farming is done by the Amish. This exhibit is a fine example of tobacco production in Maryland. (Courtesy Maryland State Fair Archives, photographed by Jason Sellman.)

An abundance of fruit is seen here. The finest peaches, apples, and pears Maryland has to offer are displayed. (Courtesy Maryland State Fair Archives.)

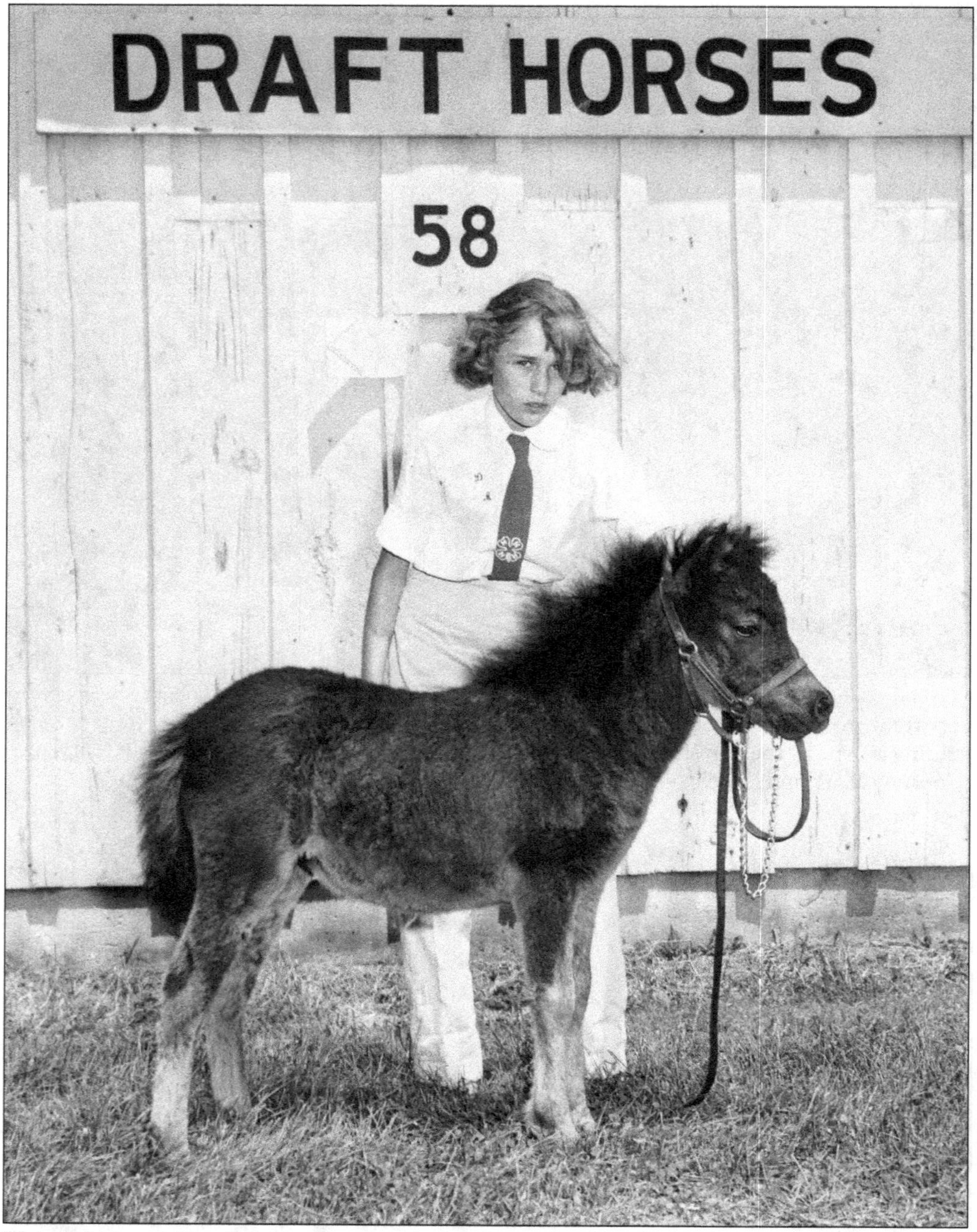

This ambitious 4-H member is photographed with an adorable Shetland pony in front of the draft horse barn. The 4-H pledge is: "I pledge my head to clearer thinking, my heart to greater loyalty, my hands to larger service, my health to better living for my club, my community, my country and my world." With the look of determination showing, there is no doubt of success. (Courtesy Maryland State Fair Archives.)

This draft horse is being shown in hand at the 1951 Maryland State Fair. The Percheron gelding being judged is standing correctly and has his mane braided in a traditional way that shows off his massive neck, complete with ribbons. Draft horse competition was very popular in the days before cars, but as the draft horses were replaced by tractors on the farm, showing them became less popular. They do, however, still have a large and devoted following. (Courtesy Bob Shirley.)

The Shirley family is synonymous with the Maryland State Fair. Their involvement goes back over 60 years, as various family members have and still participate in the fair each year. Joseph Shirley is seen here holding Bel Air Jesse James, a two-year-old Clydesdale that won Junior Champion. (Courtesy Bob Shirley.)

Joseph W. Shirley is seen driving Civic Princess, one of his many draft horses. (Courtesy Bob Shirley.)

The annual livestock parade took place on Labor Day on the racetrack in front of the fans before the start of the first race. In this photograph, Bob Shirley can be seen driving a team of his draft horses. The near horse (on the left) is Civic Princess; the far horse (on the right) is Civic Bess. Coming up from behind, leading the grand champion Clydesdale stallion, is John Aldred. (Courtesy Bob Shirley.)

Joseph Shirley is seen here again driving his draft horse Broadview Belle in the annual livestock parade. Bill Howard can be seen leading a Shetland pony. (Courtesy Bob Shirley.)

Dr. Joseph Vial (holding trophy), a livestock specialist from the University of Maryland, is presented with an award from the Maryland Draft Horse Breeders Association for his service and dedication. (Courtesy Bob Shirley.)

This 1972 photograph shows Dan Shirley receiving the Lawrence B. Sheppard Challenge Trophy from Maryland Horse Breeders Association executive director Snowden Carter. Holden Rogers, superintendent of the horse department, stands in the background at left. The Sheppard Trophy is awarded to the most helpful exhibitor in the horse department. (Courtesy Bob Shirley.)

Jim (second from right) and Bill Eckles (far right) are shown here with their prize-winning horse pulling team. Horse pulling contests are always a favorite with the crowds at the fair. The rules governing this sport are very strict. The drivers must never raise their voices to the team but encourage them with chirps, clucks, and clicks of the tongue. If the driver brandishes a whip at the team, he is disqualified. (Courtesy Bob Shirley.)

Bob Shirley (right) and Hank Parlette (center) of Baltimore County receive the 4-H Draft Horse Fitting Champion Trophy from the superintendent of the horse department, Joseph W. Shirley. When judging draft horses, the basic principle to keep in mind is what the draft horse's primary function is. His job on a farm was to pull. Farmers relied on draft horses to pull heavy farm equipment and carriages. Their massive size—often well over 1,800 pounds—makes them ideal for this type of work. (Courtesy Bob Shirley.)

This exquisite trophy of a Percheron horse was created by the artist Calvin Roy Kinstler. Kinstler grew up on a dairy farm near Baltimore. He loved to take care of and study the sturdy draft horses that were used on the farm. His love turned into a career when he began carving horses from wood. He soon realized he could not master all horse breeds and decided to concentrate on draft horses. He would make intricate sketches, from which wooden blocks of Philippine or Honduran mahogany were then carved to match. It took six days for each sculpture to be completed. (Courtesy Maryland State Archives, photographed by Jason Sellman.)

It is hard to tell if the group of teens seen in this 1959 photograph of the Midway are working at the fair or simply there to enjoy the sites. The 1950s were a time of revival and renewal for the fair. (Courtesy Baltimore County Public Library.)

After the fair resumed in 1946, fair president Matt Daiger called it the Great Exposition of Maryland. Matt Daiger said the name change was proposed because "the post war fair would be bigger and better than ever." What's better than an Esskay hot dog and a Pepsi at the fair? (Courtesy Baltimore County Public Library.)

Sightseeing trains were a popular way of seeing the fair. With the fair being held on over 90 acres, it could be exhausting just getting to the car. The army contributed to the enlargement of the fair by building two large structures that were eventually turned over to the fair and used as livestock buildings. (Courtesy Baltimore County Public Library.)

Strolling crowds and sunny weather combine for a jubilant crowd at the Midway. By the early 1940s, the Maryland State Fair was a well-established institution in Timonium and had been running for close to 60 years. However, the fair was not held for three years during World War II. The fairgrounds were used as a storage depot for engineers and a truck repair center by the U.S. Army. (Courtesy Baltimore County Public Library.)

This view of the 1958 Midway at the fair was taken from the grandstands. The Midway stretches all the way around the racetrack. (Courtesy Baltimore County Public Library.)

Crowds stroll along a muddy Midway during the 1933 Maryland State Fair. The foul weather cannot dampen the enthusiasm of the crowds. (Courtesy Baltimore County Public Library.)

The Midway, always a popular destination, was at times home to a circus. With such close proximity to the railroad, it was easy for traveling circuses and amusements to stop at the fair. Sheesley's Shows was one such traveling entertainment group. The group took over 33 flat cars. They provided entertainment such as Buckskin Ben's Wild West Show and rides such as the Ferris wheel, giant carousel, and the Whip. (Courtesy Baltimore County Public Library.)

A serious young man contemplates a purchase at the 1936 Maryland State Fair. Umbrellas, hats, and canes appear to be this vendor's specialty. (Courtesy Baltimore County Public Library.)

The quiet scene here depicts little of the storm that was brewing. In 1950, the Maryland Jockey Club held the controlling interest in the Timonium Fair Grounds. An out-of-state company that manufactured electric tools offered to buy the site from the Maryland Jockey Club for $500,000. The Maryland Jockey Club was deeply attracted to this offer, which would allow them to settle debt on Pimlico. William F. Schluderberg headed a Save Timonium committee to prevent the sale and raise the funds to buy Timonium from the Maryland Jockey Club. Matt Daiger, president of the fair, and Humphrey S. Finney, field secretary of the Maryland Horse Breeders Association, were also a part of the effort. Between 600 and 700 other people owned stock in the fair, but the Maryland Jockey Club owned the controlling interest. During the 1950 fair, $500 "Fair Bonds" were on sale to help save the fairgrounds. The money was raised and the fair lived to celebrate another day. (Courtesy Maryland State Fair Archives.)

An impressive crowd fills the stands at the fair on opening day during the 1950s Timonium horse racing season. The Timonium fair was still a standard stop on the half-mile track circuit that included other local tracks such as Hagerstown and Marlboro. (Courtesy Baltimore County Public Library.)

An aerial view taken in the 1950s shows the fairground in the upper left corner. The oval racetrack and large grandstands are on the far left, as well as the administration building. The road that runs through the center of the picture is York Road. The North Central Railroad, so important to the success of the fair throughout the years, can be seen just behind the grandstands. The fairground, as seen in this view, has expanded to over 100 acres. (Courtesy Baltimore County Public Library.)

A color guard and band march down the middle of the Midway alongside the racetrack on opening day. (Courtesy Maryland State Fair Archives.)

The officers of the Maryland State Fair gather with the color guard. (Courtesy Maryland State Fair Archives.)

Governor's Day is always a popular day that attracts many important officials to the fair. Standing at the far left is vice president and general manager John Heil. Gov. Millard Tawes stands at the center of the group wearing a dark suit. (Courtesy Maryland State Fair Archives.)

John Heil is standing left of center during a ribbon-cutting ceremony marking the opening of the fair. (Courtesy Maryland State Fair Archives.)

The 1960s were a time of growth at the Maryland State Fair. This photograph shows an exterior view of an exhibit hall being built. Larger exhibit halls offer space for more exhibitors to participate in the fair. Merchants and exhibitors mean more revenue for the fair and more interesting exhibits for those attending the fair. (Courtesy Baltimore County Public Library.)

This is the interior view of the new Farm and Garden Building. New buildings also included a new hog barn costing $15,000, measuring 352 by 28 feet, and including 83 pens. (Courtesy Baltimore County Public Library.)

In 1964, Maryland State Fair vice president and general manager John Heil stands at left as he listens to Pres. Howard Clark of the Maryland State Fair. They stand on the steps of the Administration Building. (Courtesy Baltimore County Public Library.)

In this photograph, the traditional livestock parade, which was held on Labor Day, can be seen on the racetrack. The fair no longer continues with this tradition. (Courtesy Bob Shirley.)

This view of nearby Towson is looking south on York Road from the corner of York and Joppa Road. There is a sign just before the railroad trestle advertising "the Fair in Timonium." (Courtesy Baltimore County Public Library.)

The construction of the Administration Building can be seen here during 1909. This large structure gives evidence to the fact that the Maryland State Fair was growing and prospering by leaps and bounds and was sure to have a permanent home in Timonium. What a difference the decades have made. Attendance at the turn of the century was around 40,000, and in the 1950s, it was over 200,000. Premiums paid out in 1900 were approximately $20,000, and in 1951, they were $80,000, with $14,000 going to 4-H alone. (Courtesy Baltimore County Public Library.)

In 1956, it was time again to update and renovate the Timonium Fair Grounds. Plans were made for $200,000 worth of improvements, including new jockey quarters, a secretary's office, a paddock, and grandstands that included new betting boards and a restaurant. (Courtesy Maryland State Fair Archives.)

The architect Van Rensselaer Saxe of Baltimore designed grandstands of concrete block to replace the older wooden structures. The grandstands included new and improved betting windows. When the old wooden structures were being torn down to make way for the new buildings, the debris was burned. The blaze raged out of control for several heart-stopping minutes but was quickly brought under control. (Courtesy Maryland State Fair Archives.)

This 1964 view was taken from across York Road looking into the main gates. Across the racetrack infield, the new and improved grandstands can be seen in the distance. (Courtesy Baltimore County Public Library.)

Once again the Maryland Beekeepers Association demonstrates one of the many uses for honey. The ancient art of mead making is examined at this fair demonstration. (Courtesy Maryland State Fair Archives.)

The Baltimore Gas and Electric Company (BG&E) has been very supportive of the fair for many years. In particular, BG&E has supported the youth of 4-H. Richard Ackler (right) and Betty Rush (left), representatives from BG&E, are helping a young girl with a presentation. The fair was a great opportunity for companies such as BG&E to demonstrate the latest kitchen gadgets to a large, interested audience. (Courtesy Nancy Ackler.)

Seen here is an unidentified fairgoer with one of former president Dwight D. Eisenhower's prize bulls. Ike had an avid interest in raising and showing prize Black Angus cattle. His farm, located in nearby Gettysburg, Pennsylvania, was comprised of a 139-acre farm owned by Eisenhower and 306 adjoining acres owned by partners. The Angus cattle were frequent participants at the fair in the late 1950s and early 1960s and won many championships. In 1957, the fair publicity committee decided to capitalize on this adventure and planned to give a champion heifer calf to President Eisenhower as a gift and present a $1,000 check to the calf's owner, nine-year-old Sharon Kay Staufer. Sharon had spent a lot of time caring for the calf and had become very attached to little Pearl. When she learned the calf was to be given away, she burst into tears. The press picked up on the story, which got back to President Eisenhower, who did not have the heart to take the calf. In the end, Sharon got to keep the calf, the cash, and a horse that her grandfather offered her to replace the calf. (Courtesy Maryland State Fair Archives.)

With a popular attraction such as this "Guess Your Name, Weight or Age" game in the Midway and a "Medieval Tortures" show in the background, the Midway was as popular and diverse in the 1960s as it is today. (Courtesy Baltimore County Public Library.)

A popular 1950s Midway attraction gathers curious fans. People still flocked to the fair, which was getting bigger every year. (Courtesy Baltimore County Public Library.)

In the 1960s, the fair was as popular as ever. On opening day in 1960, over 75,000 people attended the fair, which was held on close to 100 acres. John Heil was the general manager in 1960, and Daniel B. Brewster was the president in 1961. (Courtesy Baltimore County Public Library.)

Security guards escort a color guard through the Midway on opening day of the fair in 1966. Additional renovations included a new Farm and Garden Building. When the crowds got tired of the commotion on the Midway, they could enjoy the beauty and bounty of the local harvest in the new building. (Courtesy Baltimore County Library.)

This photograph of a tractor-driving contest shows what is important on a farm. The draft horses and mules have long been replaced by modern machines, and learning to drive one safely and efficiently is an essential skill. The 4-H Club helps improve this skill by hosting tractor-driving contests. (Courtesy Maryland State Fair Archives.)

Draft horses are still shown for pleasure. The draft horse's job on the farm was to pull. The horse must have a willingness to work with an eager expression. The horses in this line-up all seem to have prize-winning attitudes. (Courtesy Bob Shirley.)

At first glance this looks like a stagecoach pulled by the traditional "six in hand"—six horses used to pull a large load over a long distance. With a closer look, it appears the horses are in fact small ponies and the coach is a part of a circus company. This group may have been one of the many sideshows that appeared at the fair. (Courtesy Maryland State Fair Archives.)

Events that show calves have been in existence since the beginning of the fair, but one later addition was the calf scramble. Started in 1947 by the Angus Breeders of Maryland and the Maryland State Fair, this exciting competition gave 20 lucky 4-H and Future Farmers of America boys an opportunity to catch, halter, and lead across the finish line a precocious young calf. Successful boys were allowed to keep the calf on the condition that they bring it back and exhibit it the next year. (Courtesy Maryland State Fair Archives.)

This picture from the 1940s shows the grandstands and infield of the Timonium racetrack overflowing with spectators. This trend continued into the 1960s. The 1960s were a time of renewal and renovations for the fairgrounds. John Heil, general manager of the fair, headed the $119,000 project. Included in the renovations were a new 84-stall barn and a resurfaced track. Timonium would put these improvements to good use, as it had just acquired 12 more racing dates from the Cumberland racetrack. Horses were stabled at Timonium all year round to race and train. They could also van out to other racetracks such as Charles Town and come home in the evening. (Courtesy Maryland State Fair Archives.)

Three

These Are the Good Old Days

In 1963, a young man named Howard "Max" Mosner came to work at Timonium for the office of the Maryland State Fair and Agricultural Society, Inc. Pictured here are Max Mosner (left); Irvin Deggeller (center), the founder of Deggeller Amusements; and Bill Linton (right), a former vice president of the fair. Mr. Mosner started out at the fair office in accounting and quickly moved up to the position of assistant general manager to John Heil. Upon Mr. Heil's death in 1972, Mr. Mosner became the general manager of the fair. In 2000, he was named president and general manager. Max's contributions to the Maryland State Fair are too numerous to mention, but in 2005, he was inducted into the International Association of Fairs and Expositions (IAFE) Hall of Fame. This is an honor given to a select few with outstanding records of service and contributions to the fair industry. (Courtesy of Maryland State Fair Archives.)

In this 1971 picture, from left to right, Attorney General Bill Birch, 1971 Farm Queen Patricia Mullinix from Howard County, comptroller of the State of Maryland Louis L. Goldstein, Baltimore County executive Dale Anderson, and fair president Howard Clark stroll down the Midway. (Courtesy Maryland State Fair Archives.)

Each year at the fair, one day is "Governor's Day." This is an opportunity for the governor and other dignitaries of Maryland to stroll down the Midway, eat lunch, take a look at the exhibits, and enjoy the bounty that the state has produced. Past celebrations have included meeting with lucky exhibitors and being the recipient of special cakes baked in the governor's honor. Gov. Marvin Mandel is pictured second from left, and Comptroller Louis L. Goldstein is third from the left. (Courtesy Maryland State Fair Archives.)

Racing fan Carol Garcia is seen here getting ready for race day. She holds a copy of the program for the day's races. (Courtesy Maryland State Fair Archives.)

The Big Timonium Fair included 24 days of racing, a new development. Timonium bought 12 racing dates from Cumberland Race Track, another local half-mile track. The number of racing dates fluctuated in the 1970s from 24 days from 1970 to 1973 to 42 racing dates in 1974 to 1983. (Courtesy Maryland State Fair Archives, photographed by Jason Sellman.)

The year 1973 almost became the last year the phrase "and they're off" was to be heard at the Timonium racetrack. The Maryland Racing Commission wanted to transfer the racing dates of tracks such as Timonium to larger, one-mile tracks. Racing dates, much like liquor licenses, are given out by the state. Racing dates are the days in which the track can legally accept bets. Increasing racing dates would increase the revenue for the larger tracks and help with their financial difficulties. Unfortunately, that could have meant disastrous financial results for the Timonium racetrack. The fair itself was still very popular. Exhibitors were flocking to the fair in record numbers, but without income generated from racing, it would have been difficult for the fair to survive. (Courtesy Maryland State Fair Archives.)

Happy days were certainly to return to the Timonium fair when a Friends of the Maryland State Fair committee was formed. Maj. Gen. Edward Warfield served as head of the steering committee, which included two representatives from each county. General manager of the fair Howard ("Max") Mosner and publicist George S. Wills also had a tremendous influence on the committee. (Courtesy Maryland State Fair Archives, photographed by Jason Sellman.)

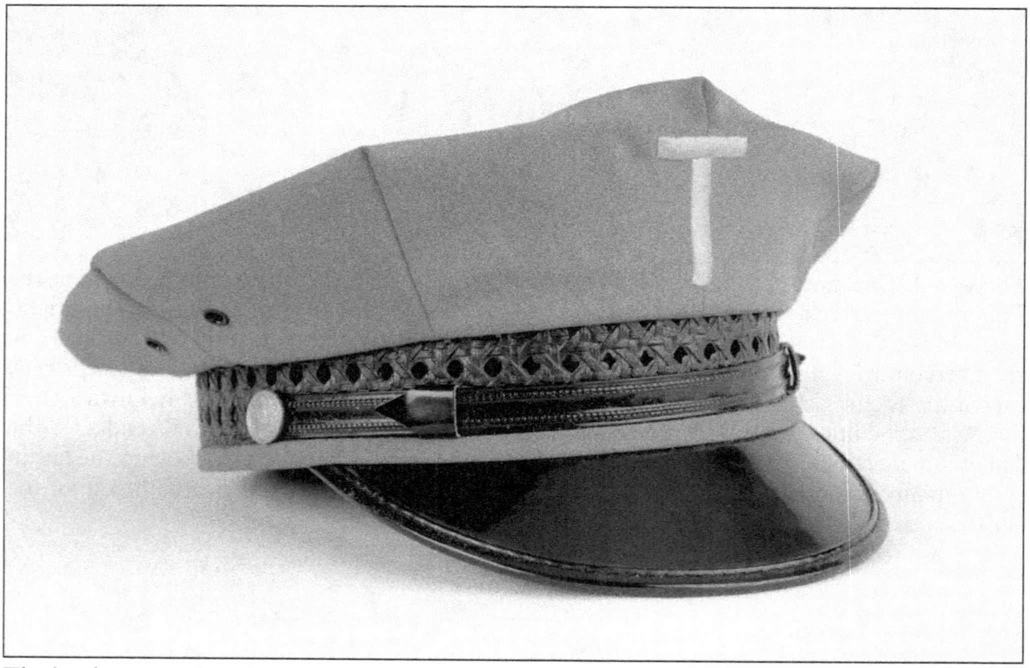

The bright orange hat shown here was worn by the ushers who worked at the Timonium racetrack. Thanks to the efforts of the Friends of the Maryland State Fair committee, ushers in the large grandstands of the racetrack could once again control the exuberant crowds, direct people to the betting windows, and make sure the racehorses remained undisturbed in the paddock. (Courtesy Maryland State Fair Archives, photographed by Jason Sellman.)

Racing fan Carol Garcia cashes in her winning ticket at the Maryland State Fair. Major General Warfield, the head of the Friends of the Maryland State Fair committee, felt that the fair and racing played an important part in the community. (Courtesy Maryland State Fair Archives.)

A gorgeous Thoroughbred is seen here either on his way to the start or to the winner's circle. (Courtesy Maryland State Fair Archives.)

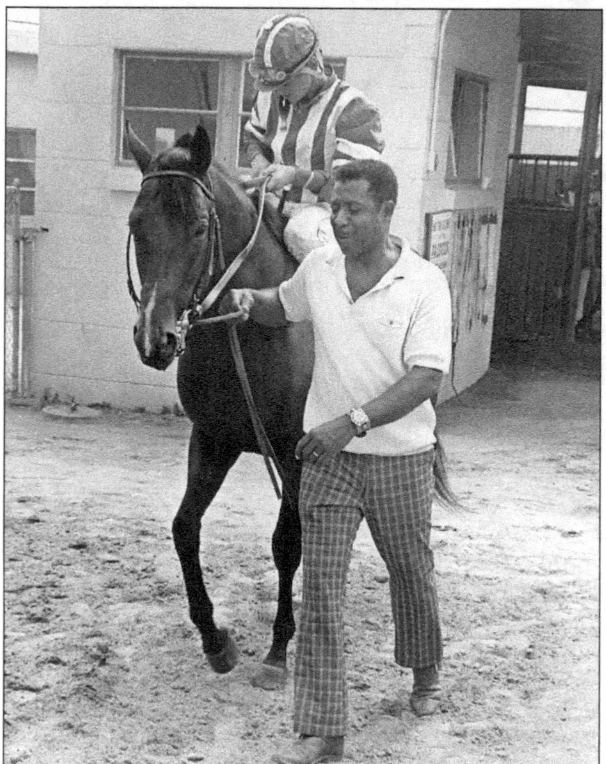

Racing was still a huge attraction to the fair in the 1970s. The fair itself, with expenses and premiums paid, would lose about $75,000. Add the $100,000 to $200,000 in revenue generated by racing, and the fair became profitable. (Courtesy Maryland State Fair Archives.)

This group of jockey, horse, and groom is just leaving the paddock on their way to the starting gate. The crowd is ready for the heart-stopping excitement of the race. (Courtesy Maryland State Fair Archives.)

Attendance at the Timonium racetrack during the 1970s averaged about 220,000 people per season, with the pari-mutuel handle at about $20 million. (Courtesy Maryland State Fair Archives.)

Throughout its history, the fair has always promised something new and exciting. In 1972, the year of this premium catalog, the fair featured stars from the country-western circuit. In 1918, the new and novel attraction was an airmail delivery. Postmistress Mary Martin received this coveted bag of mail on the infield of the racetrack. (Courtesy Maryland State Fair Archives, photographed by Jason Sellman.)

Trainer King T. Leatherbury is not only a legend at the Timonium racetrack but has attracted national attention as well. On August 23, 2003, at the seventh race of the Maryland State Fair meet, Mr. Leatherbury claimed his 6,000th victory, a feat achieved by only two other trainers in the history of horse racing. The horse Cherokee Sunrise, owned by Woodrow Marriott and ridden by jockey Mark Rosenthal, pulled away from the pack and launched Leatherbury into history. (Courtesy Jim McCue.)

When this photograph was taken, history had just been made at the Timonium racetrack—racing legend King T. Leatherbury had just won his 6,000th race. It was appropriate that Leatherbury achieved this monumental win at Timonium, as he had been racing at Timonium for 48 years and was on the Maryland State Fair Board of Directors for 11 years. He owes his success to "resilience and tenacity," attributes he claims to share with the Timonium racetrack. (Courtesy Jim McCue.)

This beautiful silver trophy is known as the Schluderberg Memorial Trophy. It was named for William Schluderberg, past president of the Maryland State Fair and head of the Save Timonium campaign in 1950 that was successful in buying the controlling shares of the fairground from the Maryland Jockey Club. (Courtesy Maryland State Fair Archives, photographed by Jason Sellman.)

The 4-H movement is still popular in 1975. David Stewart of Harford County demonstrates safe techniques for driving a tractor. (Courtesy Maryland State Fair Archives.)

Mark Abbott of Kent County is shown here with his best-in-show rabbit. He displays with pride his beautiful ribbons, large trophy, and prized rabbit. (Courtesy Maryland State Fair Archives.)

Patty Payneter of Baltimore County is seen here in front of her champion display on childcare. (Courtesy Maryland State Fair Archives.)

The 4-H Club is part of the Cooperative Extension Service, a nonprofit program operated through each state's land grant university. The extension service's staff operates 4-H offices throughout the counties of each state. (Courtesy Maryland State Fair Archives.)

Susan Harce, 1972 Farm Queen, is not just another pretty face. The competition was started in the early 1940s. The girls, ages 16 to 19, compete first for a county title and then proceed to the Maryland State Fair, where they are tested on their knowledge of agriculture, speaking ability, and personality. The winner promotes agriculture throughout her yearlong reign at a variety of events. (Courtesy Maryland State Fair Archives.)

Seen from left to right are Miss Montgomery County Gwen Hilton, Miss Howard County Kim Sullivan, Miss Allegany County Anne Denise Heavner, Miss Kent County Lee Hill, and Miss Baltimore County Ginny Ellen Wisner of 1974. (Courtesy Maryland State Fair Archives.)

The 1973 Farm Queen Jocelyn Price (center) and her court are pictured with, from left to right, (second row) Maryland State Fair chairman of the board Dr. J. Morris Reese, Pres. Howard Clark, and general manager Howard "Max" Mosner.

Lee Hill of Kent County is seen here having just been crowned Miss Maryland Farm Queen 1974. The contest stresses not only poise and beauty but farm, home, and community responsibility as well. (Courtesy Maryland State Fair Archives.)

Max Mosner (far left) and Paul Hampshire, chairman of the board (far right), are seen here with all the county representatives for the Maryland Farm Queen Contest in 1974. The girls went on to compete for the title of queen; one queen and four runners-up were selected from the group. They were judged on personality, appearance, home responsibility, and community activity. In 1974, the Maryland Farm Queen, Lee Hill, was awarded a $200 scholarship to attend the University of Maryland; runners-up received $100 and $50. Participants who were not awarded a title received $25 and a commemorative bracelet. (Courtesy Maryland State Fair Archives.)

General manager Max Mosner can be seen here escorting the 1978 Maryland Farm Queen and her court in a beautiful carriage drawn by two draft horses. The coachman is turned out in traditional livery with a top hat, high-topped boots, trimmed driving apron, and gloves. It became popular in the mid-1970s for the queen and her court to participate in the 4-H parade held on the racetrack in front of the grandstands on opening day. (Courtesy Maryland State Fair Archives.)

Showing the fruit of much labor, many beautiful examples of needlework and rug hooking can be seen in the Home Arts Building at the fair. Amateurs work hard all year to produce unique pieces that will be judged by a panel of experts. The pieces are closely examined, and ribbons are awarded. In the early 1920s, Charles Pratt of Philadelphia caused an uproar when he entered a quilt in the fair and won. Quilting was still regarded as work for women. (Courtesy Baltimore County Public Library.)

Cake baking and decorating is always a popular class at the fair. There are several different divisions one can enter, including divisions for children and teens. A favorite division, seen here, is the fair president's cake. This cake can be any kind of cake decorated in themes of the state fair. The first-prize-winning cake is then presented to the president of the fair. (Courtesy Maryland State Fair Archives.)

When people have had a special relationship with the fair, whether as life-long exhibitors or through many years of service, trophies or awards may be given in their honor. This collection of ribbons and memorabilia represents the Helen Burns Smyth Award. This award is given out by friends and family of Helen Burns Smyth to the champion exhibitor in the Home Arts, Farm and Garden, and Equestrian categories, which she loved to participate in, to honor her memory. (Courtesy Maryland State Fair, photographed by Jason Sellman.)

The Maryland State Fair celebrated 100 years in 1981 with much fanfare and reminiscing. The fair was still as popular as ever with over 500,000 visitors in 1981 and profits reaching $50,000. The controversy that seemed to follow the fair was ever-present in 1981 when Gov. Harry Hughes suggested taking the racing dates from the Timonium racetrack as well as the Bowie Race Track to make Laurel and Pimlico the only racetracks in Maryland. (Courtesy Maryland State Fair Archives, photographed by Jason Sellman.)

Howard Hurline from Baltimore County works hard to take care of his cows. The 4-H Club teaches youths valuable lessons such as responsibility. (Courtesy Baltimore County Public Library.)

Swine have always been a favorite with the spectators at the Maryland State Fair. Wendy Anne Smith (left) and Jessica and Emily Hester look on as the pigs enjoy a meal. The animal on the left is a Duroc, and the one on the right is a crossbred, both very popular breeds. They are judged on a number of points such as top line, neatly carried bottom line, and overall appearance. (Courtesy Baltimore County Public Library.)

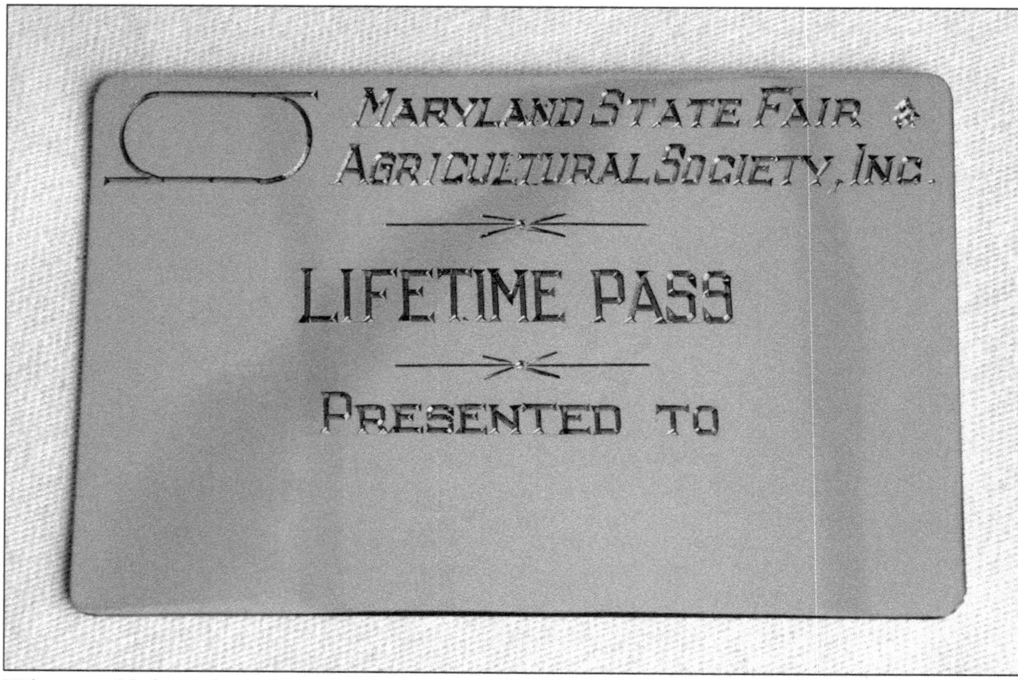

What would the ticket takers think if presented with this silver Lifetime Pass? The layout of the racetrack can be seen in the upper left-hand corner. (Courtesy Maryland State Fair Archives, photographed by Jason Sellman.)

From the old to the new—this program from 1889 was only 20¢. Compare this with the program on the following page. (Courtesy Maryland State Fair Archives photographed by Jason Sellman.)

The beginning of the new millennium was celebrated at the Maryland State Fair, still going as strong and as popular as ever. The goal in the 21st century is still to educate and entertain all who flock to the fair. (Courtesy Maryland State Fair Archives, photographed by Jason Sellman.)

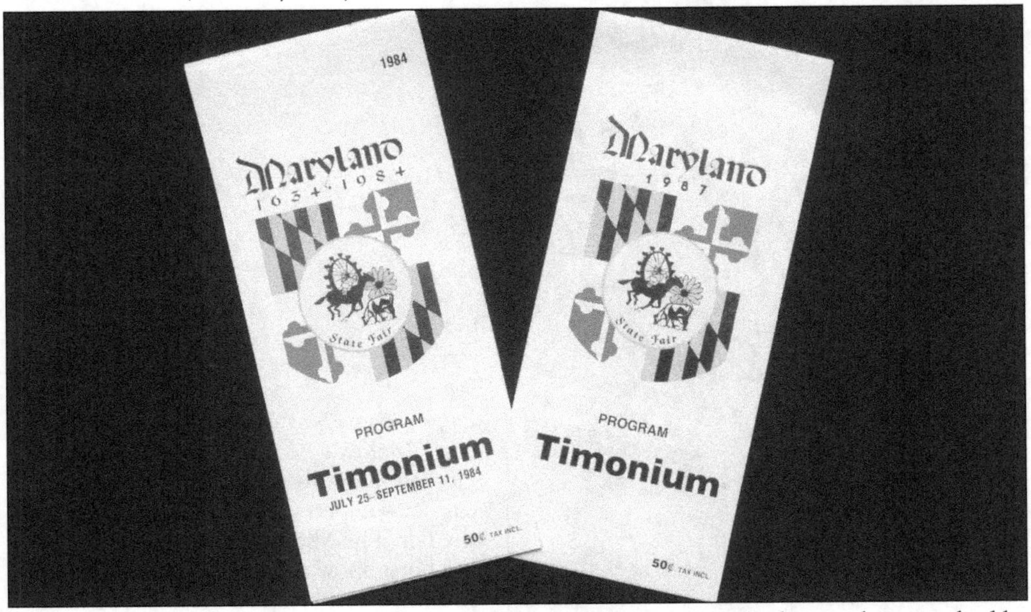

This program from 1984 is only 50¢. (Courtesy Maryland State Fair Archives, photographed by Jason Sellman.)

Paige Horine has been competing at the Maryland State Fair since 1976, and she has entered everything from peanut butter fudge to needlepoint to a Spam quiche. Now the Horine family joke is "we enter the children." Paige and daughter Devon are seen here just before Devon goes in to ride for the first time by herself at the state fair. (Courtesy Edie Bernier.)

Paige Horine is seen leading the family pony, McKay's Aloha, while Amelia Horine enjoys the ride. The muddy ring cannot dampen the enthusiasm and fun of the fair. The Midway can be seen behind the horse show ring. Only the quietest and most well-behaved horses and ponies can stand the noise and commotion of the Midway. (Courtesy author's collection.)

The Maryland State Fair has always been an event enjoyed by families. The family of Paige Horine turns out to cheer on Devon, riding in a lead-line class. Pictured from left to right are Daphne Grosh, Paige Horine's sister; Paige's nephew, making faces for the camera; Paige Horine (center) holding her daughter Amelia; Paige's mother, Doris Rolfes, standing ready to catch a loose pony or child; and Paige's aunt Joanne Weber. Devon sits calmly on her pony, McKay's Aloha, happy with her fourth-place ribbon. (Courtesy author's collection.)

Paige, daughter Amelia, and pony Aloha are seen again getting ready to line up to be awarded a ribbon. Amelia is having so much fun she forgets to hold on to the reins with both hands, preferring to wave to her family on the outside of the ring. (Courtesy author's collection.)

The Midway, with its glorious view, is a prime destination for the young and old. Deggeller Attractions is the company that puts on the Midway at the fair. They have been supplying the fair with rides and games since 1979. The company was started by Allen and Irvin Deggeller and is still run by the Deggeller family. Every single ride and game of skill is portable and can be packed away to transport to the next destination. Over $2 million worth of rides and games are a part of the Midway, and they travel from state to state 10 months out of the year. (Courtesy Edie Bernier.)

The Ferris wheel is one of the most popular and oldest rides on the Midway. The Ferris wheel first made its appearance at the fair in the early 20th century. (Courtesy Edie Bernier.)

Everyone has his or her favorite rides on the Midway. The swing seen here always attracts a crowd. (Courtesy Edie Bernier.)

The fair generally opens at 10:00 a.m., with livestock shows starting closer to 9:00 a.m. Those exhibiting livestock often get to the fairgrounds before dawn, as it takes a long time to properly groom an animal for the show ring. When the Midway opens, the fair really comes alive. (Courtesy Edie Bernier.)

What started as just a one-day event has become by 2006 the "Eleven Best Days of Summer." Over the years, the length of the fair has fluctuated, often to mirror racing dates. In 1954, it was the 14 best days of summer. (Courtesy Edie Bernier.)

If all the excitement of the rides leaves one feeling hungry, there is never a shortage of food at the fair. Maryland specialties are featured at the McCormick Maryland Foods Pavilion. Crab cakes, catfish filet sandwiches, fried clams, turkey legs, corn on the cob, and much more can be found at the base of the Midway. (Courtesy Edie Bernier.)

Nothing beats the thrill of the thunder of hooves as a field of Thoroughbreds passes in front of the grandstands on their way to the finish line. The horses are so close, viewers may get mud splattered on them from a wet track. But if it is a "dark" day—a day without live racing—fans can always come to the track and enjoy a simulcast race from several other racetracks around the country. (Courtesy Edie Bernier.)

The glow of a balloon as it ascends into the night is an awesome sight. Balloon ascensions and hot air balloon races were a popular attraction at the fair in the 1990s, but balloon races at the fair were stopped due to security reasons after September 11, 2001. Balloon glows, or tethered balloons, can still be seen at the fair. (Courtesy Edie Bernier.)

This balloon gets ready to take off on a glorious day fit for adventure. The largest balloon race held by the fair featured over 20 balloons. (Courtesy Edie Bernier.)

Many hot air balloons were sponsored by companies as a means of advertising. Who could miss a giant bottle of champagne or the Energizer Bunny in the air? (Courtesy Edie Bernier.)

This gorgeous bushel of apples promises a lot of delicious pies. There are several different classes in which an exhibitor can enter apples depending on the variety. Familiar varieties of apples include Granny Smith, McIntosh, and Red Delicious, but Stayman and Summer Rambo can also be seen on display. (Courtesy Edie Bernier.)

A bountiful harvest is seen in this lovely display. The fair at the beginning of the 21st century still holds true to its agricultural roots and in fact is still run by the Maryland State Fair and Agricultural Society, Inc. (Courtesy Edie Bernier.)

A cornucopia of vegetables is seen, complete with a champion ribbon. A blue ribbon is nice, but a purple rosette is best—a blue ribbon signifies first place, and a purple rosette signifies champion. (Courtesy Edie Bernier.)

This basket of vegetables looks very much like one that a local farmer may have brought to the fair over 125 years ago. This basket probably traveled to the fair in an air-conditioned car as opposed to horse-drawn carriage. One thing is the same: the vegetables still flourish with careful attention, and the farmers who grew them were equally proud. (Courtesy Edie Bernier.)

The Zipper, the most ferocious ride on the Midway, is not for the faint of heart. This colossal ride spins and twirls until the riders are dizzy. This is a far cry from the Ferris wheel and the Whip that were so popular in the early 20th century. The Midway is a beautiful sight at night. (Courtesy Edie Bernier.)

When you are finished riding the rides, if your stomach allows, there are always enough food choices for the most discriminating palette. The traditional cotton candy and candy apples are at the Midway, along with pizza, hot dogs, and funnel cakes. (Courtesy Edie Bernier.)

Some children enjoy the Midway more than their parents. (Courtesy Maryland State Fair Archives.)

The Midway has something for everyone. Children can enjoy special rides that cater to just their size and speed. (Courtesy Edie Bernier.)

The competition is stiff among Maryland's finest produce. According to the 2003 *Premium Catalog*, the basis for judging vegetables is as follows: "Exhibits will be judged on the basis of quality, condition, uniformity, and trueness to variety. It should be free from blemishes caused by disease, insects, or other factors. The average size from the commercial standpoint, rather than the largest specimen should be exhibited." (Courtesy Edie Bernier.)

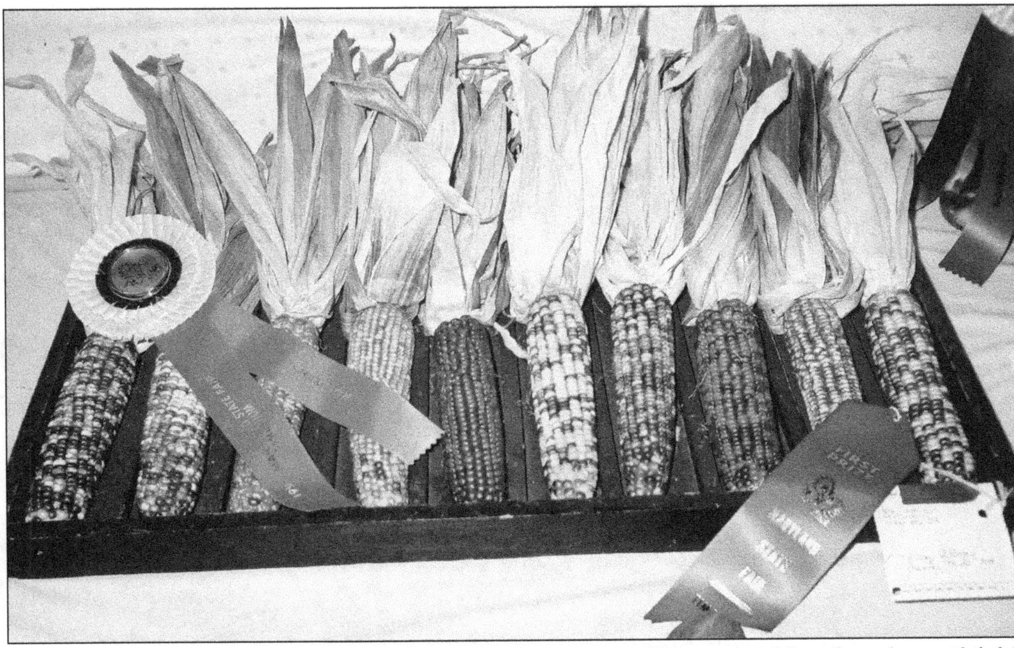

This harvest of corn, proudly displaying several winning ribbons, would make a beautiful fall centerpiece; however, this display was probably entered in the 10 ears feed corn class. The judges are looking for excellent "ear and stalk quality, freedom from disease and insect damage." Because corn is harvested closer to the fall, this display is from the crop harvested the year before. (Courtesy Edie Bernier.)

The ribbon shown here, with its yellow rosette, gold center, and yellow tails, was at one time the symbol of a champion. In the center of the rosette is the seal of the state of Maryland. This ribbon was used in the early 1990s and then discontinued for the large, solid purple ribbons used in the 21st century. (Courtesy Edie Bernier.)

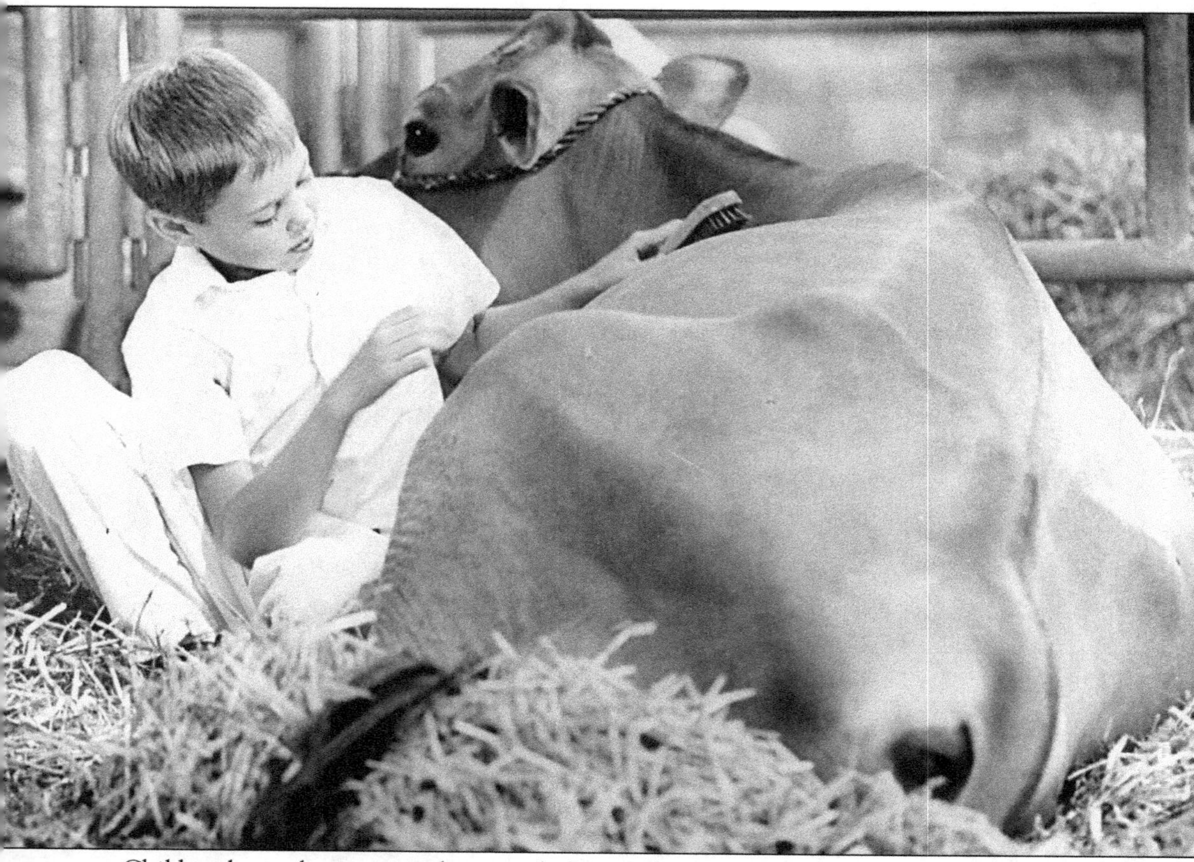

Children have always enjoyed a special place at the fair. It is a place they come to play on the rides, eat cotton candy, and win a prize. The fair is often the first place a child gets to sink their hands into the woolly fleece of a lamb, to feel how scratchy a pig's back really is, or to see where milk comes from by milking the cow themselves. Children also have the opportunity to learn important life lessons in competition. They can compete against adults on equal footing and receive a cash prize if they win. The responsibility for the care of an animal that they have on a farm, day in and day out, comes to fruition at the fair in front of their peers when a panel of experts looks at their animal and says "well done." The fair is a year-long goal that the children work toward. They often think about the fair with every stitch, early morning milking, or recipe that did not turn out. It's a goal that they work toward with the determination and single-mindedness that belies their youth. (Courtesy Edie Bernier.)

The Maryland State Fair means many things to the millions of people who have displayed at or ever attended the fair. It is a tradition that, at its roots, has continued very much in the same way it did 125 years ago. The Maryland State Fair belongs to the people of Maryland, whose love of their state shows every year in every exhibit with only the kind of pride and support that Marylanders can show. (Courtesy Edie Bernier.)

www.ingramcontent.com/pod-product-compliance
Lightning Source LLC
LaVergne TN
LVHW081933200926
848484LV00009B/2415

9781531625924